Essays on Labor Market and Human Capital –
Korea and Germany

European University Studies

Europäische Hochschulschriften
Publications Universitaires Européennes

Series V
Economics and Management

Reihe V Série V
Volks- und Betriebswirtschaft
Sciences économiques, gestion d'entreprise

Vol./Bd. 3383

PETER LANG
Frankfurt am Main · Berlin · Bern · Bruxelles · New York · Oxford · Wien

Mee-Kyung Jung

Essays on Labor Market and Human Capital – Korea and Germany

PETER LANG
Internationaler Verlag der Wissenschaften

Bibliographic Information published by the Deutsche Nationalbibliothek
The Deutsche Nationalbibliothek lists this publication in the Deutsche Nationalbibliografie; detailed bibliographic data is available in the internet at http://dnb.d-nb.de.

Zugl.: Frankfurt (Main), Univ., Diss., 2010

D 30
ISSN 0531-7339
ISBN 978-3-631-61799-1

www.peterlang.de

To
my mother
In-Ho Choi

Preface

'Essays on Labor Market and Human Capital – Korea and Germany' is a doctoral dissertation at the Faculty of Economics and Business Administration of the Johann Wolfgang Goethe-University in Frankfurt am Main. This dissertation was mainly based on in critical reviews concerning human capital investment theory and labor markets in the seminar of 'Arbeitsmarktökonomik' by Professor Roland Eisen. The seminar on 'Labor Economics' by Professor Bernd Fitzenberger provided a link between my critical study of the human capital investment theory and the econometric methodology. Benefitting from these lectures, I developed the doctoral dissertation project, which endeavors to identify major reasons and solutions regarding contradictions between the problematic situation of human capital investment in Korea and the theory of Becker's human capital investment (1962). While conducting this doctoral dissertation project, advice by Professor Horst Entorf was crucial to my understanding of the theory and methods of 'Micro Econometrics' and to developing competence in the application of econometric models. His guidance was also the key to the results of my work.

Publishing this doctoral dissertation, I would first of all like to thank my supervising Professor Roland Eisen for his support and encouragement. His full confidence concerning the progress of my work enabled me to complete this dissertation project successfully. I am also grateful to Professor Horst Entorf for his valuable suggestions and advice on the application of econometric techniques. Moreover, I take the opportunity to thank Professor Bernd Fitzenberger for his seminar. I further benefitted from discussions and insights at the doctoral seminar on 'Applied Econometrics' and the seminar on 'Development and Transition' at Frankfurt University, and the workshops at the Institute for Employment Research (IAB). Additionally, I thank Mike, Hyun-Kyong, Seok-Soon and Seung-Hyun for proofreading my doctoral thesis and for providing helpful remarks.

The principal objective and contribution of my dissertation is the comparison of the returns on human capital investment in Korea and Germany, utilizing empirical methodology. This econometric comparison is a novelty, not only in the area of labor economics but also in the general field of economic studies. My skills and competence in this comparative study were enhanced by taking part in a one-year further educational course in labor issues at the 'Akademie der Arbeit' at Frankfurt University, which primarily focused on Politics, Labor and Constitutional Law, Social Welfare System, Economic Policy as well as on Sociology and Philosophy; this was followed by my diploma-oriented studies in Economics at the J. W. Goethe-University in Frankfurt am Main. These studies offered the chance to profoundly understand German society, especially its (vocational) education system. I am really grateful to Hans-Böckler-Stiftung for their scholarship forstering my further educational training programs and diploma.

Finally, I owe thanks to my father and mother for their unlimited love and efforts to educate their children; I also thank Yun-Rok, Yeon-Bo, Hyun-Soon, Hyo-Young, Jin-Ho, Yu-Chang and Seung-Jin for their encouragement. The publication of the doctoral dissertation is supported by Mr. Dong-Hoe Jung and Mr. Bong-Hyun Son. I greatly appreciate their financial support in publishing my dissertation.

Frankfurt am Main, March 1, 2011 Mee-Kyung Jung

Contents

Tables

Figures

General Introduction

Excessive competition for university entrance is a very serious problem in Korea[1]. Since 2004, more than 80% of all high school graduates have gone on to university or junior college. In 2008, this reached 83.8%, although only 76.7% of graduates from all tertiary educational institutions found a job. Only 56.1% of all graduates have a regular job (Korean Educational Development Institute (2009)). The percentage of the unemployed who graduated from a university was 12.6% among all the registered unemployed in 2000, 17.1% in 2004, 19.5% in 2007 and 22.04% in 2009 (Korean Statistical Information Service (2010)). In spite of this fact that more highly educated people suffer more seriously from unemployment, the desire to go to university still keeps growing in Korea. On the other hand, the number of vocational high school graduates continues to fall. Job training is generally neglected in Korea.

In human capital theory, as developed by Becker (1962, 1964), a positive correlation between earnings and the level of skills – school education and on-the-job training – and the reduction of the unemployment rate with the increasing level of skills was determined to be the stylized fact of human capital investment. However, the current situation in the Korean labor market does not serve to justify this theory of human capital. In studying the contradiction between theory and reality in Korea, this study seeks to uncover some reasons behind the biased behavior of Koreans concerning human capital investment, which focuses too much on university education. Starting from this critical mind-set, this study further compares the situation in Korea with that in Germany in order to find a possible solution, which could lead to a balance between labor supply and -demand concerning university graduates and graduates from vocational high schools. With this in mind, this study seeks to determine the effects of German vocational education and training.

Finally, this study focuses on human capital investment and women's earnings. Korea and Germany share the common problem of a low female fertility rate. In the theory of human capital for women as developed by Mincer und Ploachek (1974), childbirth and childcare are recognized as a career hindrance for working women. With the growing educational level of women, women's ambition in working life becomes higher. However, the social progress concerning family life lags far behind women's enhanced self-consciousness. This disparity could be reflected in the low fertility rate. This study tries to find the effect of having

1 The national description is written as Korea instead of South Korea in this study.

children on women's careers, as measured by their earnings, and finally compares the difficulties of having children for working women in Korea and Germany.

This dissertation is comprised of the following three empirical studies: The first essay is entitled "*What can Job Training do? Is University the Sole Way of Life? - The Effects of School Education and Job Training on Wages in Korea.*" This study, in the framework of human capital investment theory (Becker (1962, 1964), Mincer (1974)), attempts to find the reasons behind the overheated fever surrounding university entrance in Korea, while simultaneously neglecting job training. Although the theory of human capital has already pointed out that job training is directly oriented to increasing productivity, unlike school education, the significance of job training in human capital investment is almost entirely neglected in Korea. Why does Korea invest so much in school/university education while neglecting job related education and training? Concerning this question, a possible hypothesis would be as follows: Korea gives a great deal of weight to school and university education, because the effect of graduating university on wages is much greater than wage effects associated with other lower educational levels. This wage difference, however, can not be compensated for with any other human capital investment in Korea. Job training, which generally enhances one's productivity and earnings parallel with formal education, has only a marginal influence on wages. So, Koreans do not recognize job training as an alternative factor in human capital investment. This is the primary reason why people invest exclusively in school and university education and consider job training to be of no account. To test the hypothesis, this study estimates different wage effects between educational levels and job training and to this end, compares the effect of university education with that of job training, which comprises a new project in Korea.

Concerning the educational system, the present school system in Korea was influenced by the USA and was established in 1949. The school system mainly consists of 6 years of elementary school (comparable to *Grundschule* in Germany: ISCED-97[2] level 1), 3 years of junior high school (*Real-/Hauptschule* or *Gymnasium*: ISCED-97 level 2), and 3 years of senior high school (*Berufsschule*

2 ISCED-97 stands for the International Standard Classification of Education 1997 by UNESCO. ISCED is a framework to collect and report data on educational programs varying widely between countries with a similar level of education. It helps with the compilation of internationally comparable education statistics and indicators. See OECD (1999).

or *Gymnasiumoberstufe*: ISCED-97 level 3 and 4). At the tertiary level (ISCED-97 level 5 and 6), there are 2-3 years of junior college (*Fachschule* or *Meisterschule* in Germany) or 4 years of bachelor's studies. It proceeds then with a 2 years master's course and a further 3 years or more for the doctoral course. Senior high school is divided into vocational high school (such as *Berufsschule*) and academic high school (such as *Gymnasiumoberstufe*). Vocational senior high school in Korea even teaches pupils theories concerning vocational specialties between the 11th and 12th classes, but this school does not carry out vocational training, which would enable pupils to acquire vocational qualifications. After finishing school education, job training is generally carried out by the company, the government or on one's own initiative.

In order to estimate the effects of school education and job training on wages, this study employs the Mincer earnings model. The model, however, reveals some problems in estimating the effects of school education and job training on wages by means of ordinary least squares (OLS). i.e. measurement errors associated with the educational level, unobserved omitted ability bias and self-selection bias concerning schooling and job training, and sample selection bias on wage data in this study. Ashenfelter and Zimmerman's studies (1993, 1997) demonstrated that omitted ability and the oppositely directed measurement error offset their respective biases in a sample of the same size. Concerning the remaining selection biases, this study makes a couple of assumptions about the Korean data deduced from the particular educational and social circumstances in Korea. These assumptions enable us to compare educational effect on wages with the job training effect, although the job training variable still suffers from omitted ability and selection biases. The estimation methods utilized in this essay are the random effects method and the Heckman 2 step method (Heckman (1979)). An advantage of the random effects method is its competence, which removes serial correlations between unobserved individual effects and idiosyncratic errors in the panel data. The Heckman 2 step method corrects bias due to sample selection.

Data, used in this study, is taken from the 'Korean Labor and Income Panel Study (KLIPS)' which is comparable to the 'German Socio-Economic Panel SOEP (GSOEP).' This survey was launched in 1998 with 5,000 households and their 13,321 family members, and composed of employees, self-employed workers, non-employed people and members of the economically inactive population. From this data between 2002 and 2006, this study uses 6,649 observations for males aged between 26 and 55. The dependent variable of this study is the natural log of individual monthly net wages from their main job. The main explanatory variables are levels of school education and job training experience in employees' whole working life. Other control variables are 10 labor market experience dummies, 10 tenure dummies, 5 year dummies, 16 dummies of residence of house, 202 industry dummies, 19 firm type dummies, 8 firm size dummies, 5 casual worker dummies, and dummy variables for regular worker, part time, overtime, union, shift work, household head and having children. Additionally, this study employed 4 house ownership dummies, natural log household sustenance allowance for parents, natural log non household labor income, natural log household financial wealth and natural log household debt in order to correct the sample selection bias in the estimation obtained via the Heckman 2 step. At the end of the essay, this study attempts to bring to light whether job training could compensate or noticeably reduce the wage difference attributable to different educational levels. This study found a small wage effect due to job training. Koreans may not be aware of job training as an alternative factor in human capital investment.

In the second essay, this study conducts a comparative analysis. The subject of the study is "*Effects of School Education and Job Training on Earnings in Korea and Germany. A Comparative Study*." The central question of this essay is: why is a university education in Korea considered to be so much more important than is the case in Germany? In attempting to answer this question, Korean researchers often advance the hypothesis that the earnings difference between university graduates and workers with a lower level of education is larger in Korea than in Germany, where vocational training at school and on-the-job training for adults are well developed. This job training reduces earnings differences between educational levels and ameliorates inefficient competition for university entrance (Kim (2006)). This study estimates effects attributable to school education and job training on earnings and compares these effects between Korea and Germany. A comparative study, in which the earnings effects of human capital are estimated and compared in the same empirical framework, is a new endeavor in the study of human capital investment, but also in other thematically related areas of empirical study in Korea. This comprises the main contribution of this study.

Job training in Germany is primarily carried out at vocational school (*Berufsschule*) in the dual system. In this system, pupils take part in vocational training for 3-4 days a week in a firm. For an additional 1-2 days, pupils study vocational oriented theories at school. Firms which conduct vocational training are responsible for the vocational qualification of pupils, corresponding to regulated qualification standards. At the end of their training, these pupils have to pass an examination to demonstrate their practical and theoretical competence in order to finish the training course. After successful completion of training, they obtain a certificate, which serves as their vocational qualification. In 2004, 52.5% of labor market entrants in the relevant age group took an accredited job, which is acquired at vocational school in the dual system. 20% of matriculating students had vocational training in 2005. Graduates of vocational school have a further chance of acquiring additional vocational qualifications at a *Fachschule* or a *Meisterschule* – i.e. the advanced vocational school - when they gain practical experience in an appropriate job according to the rules (Hippach-Schneider et al. (2007)).

In Korea, 3 years of vocational senior high school does not provide a job oriented qualification program for pupils. Even if they could participate in vocational practice in a firm, this practice does not furnish pupils with any accredited vocational qualification. This, as a rule, takes only 34 hours to 6 months. There is no standard curriculum for the practice and qualification exam or certifications, which demonstrate pupils' vocational qualification. For all senior high school graduates, the percentage of vocational high school graduates is 31%. In 2004, only 11% of new comers in the labor market were graduates from vocational senior high school (Korean Educational Development Institute (2005), pp. 31). From the view-point of labor supply, the vocational senior high school does not play a significant role. In general, job training is mainly carried out by firms and the government, or individuals voluntarily take part in any job training for one's own qualification. The proportion of participants averages about 5% of the entire population aged 16 or older.

This second essay, based on the Mincer earnings model (1974), employs the quantile regression method (Koenker and Bassett (1978), Koenker (2005)) for estimating school and job trainings effects on earnings in Korea and Germany. Quantile regression enables us to estimate the effects of education and job training at different distributional points of log earnings, which are taken to be the 10th, 25th, 50th, 75th and 90th quantile points in this study. Thus, we are able to analyze not only earnings differences between educational levels, but also the differences between various distributional points for a particular level. In a comparative study, this advantage especially expands the comparability manifold. The quantile regression method allows us to measure earnings differences attributable to the variables of school education and job training throughout the whole distribution in Korea compared to those in Germany as well as to look at the effect of workers' heterogeneity on the earnings between both countries. This essay uses the 2002 to 2007 data for male employees aged between 18 and

55 taken from the 'Korea Labor and Income Panel Study (KLIPS)' and the 'German Socio-Economic Panel (GSOEP).' GSOEP was launched in 1984 and surveyed 5.921 households and their 12,290 family members aged 16 and older. The dependent variables of this study are log yearly earnings of employees in both countries. The independent variables are mainly educational levels/duration and indicator of completed job training in the last a year. Using these variables, this study utilizes 7,359 observations for 2,709 individuals from the KLIPS und 13,011 observations for 4,129 individuals from the GSOEP. According to estimation results, this study falsifies the initial hypothesis. Earning differentials between education levels are higher in Germany than in Korea, although German job training compensates for these differentials more strongly than Korean job training.

This thesis then turns its attention to working women with/without children and their careers, as measured by their earnings, in the third essay. Since 2005, Korea has recorded the lowest fertility rate across all OECD countries. This fertility rate equaled only 1.08 babies per woman aged between 15 and 49 in 2005. In Germany, although the fertility rate has not declined dramatically, it has remained below the natural replacement rate of 2.1 babies since the 1970's. Entitled "*Women's Career and Children -Human Capital and Earnings of Women in Korea and Germany. A Comparative Study*," this essay seeks to study the impact of childbirth and childcare on the earnings of women. Following the study concerning human capital and women by Mincer and Polachek (1974, 1978), this thesis advances the hypothesis of earnings discontinuities suffered by working mothers due to problems associated with childbirth and childcare. This study expected an earnings decrease for working women due to childbirth and childcare. The purpose of this study is to identify any such earnings effects linked to having children for women in Korea and Germany. As the first comparative study concerning human capital and women's earnings, this study also plans to demonstrate the effects of education, experience and family characteristics on earnings in both countries. In Korea, research on the human capital of women and the subject of children and their effect on women's earnings remains an undeveloped area of study. A comparison of human capital effects on women's earnings in Korea with similar effects in Germany is, in general, a completely new undertaking, which will contribute to the study of human capital research and the labor markets in Korea.

Observing the participation rates in the higher level of education, the female entry rate for a junior college averages 53% based on each year of relevant age (i.e. 17% OECD on average), and the entry rate for university amounts to 59% (63%

OECD). The employment rate of Korean women between the ages of 25 and 64 was 58.3% (64.9% OECD) in 2007. Compared to the OECD average (79.5%), women with a tertiary education had much lower employment rates in Korea (61.2%), while women with an education level below upper secondary had much higher employment rates (58%) compared to the OECD average (47.8%) for this group of women in 2007. Korea is therefore only an exceptional case among all OECD countries, where women who attain a tertiary level of education generally have a higher participation rate in the labor market than women with lower educational levels. Average annual earnings for females as a percentage of those of males aged between 25 and 64 equal 65% for women with a tertiary education. This amounts to 47% for women with an upper secondary educational level. One third of working women have a precarious working contract. Public support in monetary payments stands at 0.175% of GDP, which averages 2.4% for 24 OECD countries (OECD (2007)). Women aged between 15 and 49 had just 1.08 babies on average in 2005. This fertility rate is the lowest of 30 OECD countries in 2005 (OECD (2009b)).

In Germany, the female entry rate to institutes for higher vocational qualification such as *Fachschule* or *Meisterschule* averaged 16% based on each year of relevant age for women in 2007. The entry rate to *Fachhochschule* and university was 35% for females. The employment rates for females between the ages of 25 and 64 were high at 67.3% for all educational levels in 2007. German women who have attained a tertiary level of education had an average participation rate of 80.6% in the labor market. This participation rate is higher than the rate for women with an education level below upper secondary and an education level of upper secondary. Average annual earnings for German women as a percentage of men's earnings were 59% for holders of a tertiary education degree in 2007. This percentage was also 59% for women with an education level of upper secondary (OECD (2009a)). Public support exclusively for families was equal to a total of 3.0% of GDP in 2003 (OECD (2007)). This level is higher than the 2.4 % OECD average. However, formal childcare for children younger than three years old is especially inadequate. The fertility rate in Germany (1.34 babies in 2005) is higher than Korea. But since the 1970's, the fertility rate has fallen below the mortality rate (OECD (2009b)).

The model by Mincer and Polachek (1974), on which this study is based, divides the labor market experience of women into three phases – employment phase before marriage, home time for childcare and employment phase after home time, which differentiates the human capital earnings model for women with children from the model for men or women without children. By means of quantile regression, this study demonstrates, in a general fashion, human capital effects on women's earnings. Regression discontinuity design method (Thistlethwaite and Campbell (1960)) contributes to the estimation of effects of having a child or two children on women's earnings and further identifies any earnings discontinuity for women due to childbirth and childcare along the tenure earnings function, which is a main interest of this study. For this estimation, the data of females aged between 30 and 55 is taken from the 'Korean Labor and Income Panel Study (KLIPS)' and the 'German Socio-Economic Panel (GSOEP)' between 2002 and 2007. This study makes use of 3,066 observations in Korea and 8,260 observations in Germany. The main variables utilized in this study are the log gross yearly labor earnings of individuals as dependent variable, and age, tenure, dummies for marital status, single mother, household head, children number and educational levels/years as the independent variables. This study found earnings discontinuities for women aged between 30 and 40 in Germany and women aged between 40 and 55 in Korea. The German women suffer from a more than 60% earnings drop due to having children at an average tenure of 7 years. Unexpectedly, the Korean women show a 37% earnings gain attributable to children at an average tenure of 4 years.

The main part of this doctoral thesis begins with an analysis of the human capital effect on wages in Korea, focusing on school education and job training. Thereafter, this thesis compares the influence of school education and job training on earnings in Korea with the situation in Germany. Then, I focus on the subject of the human capital of women, primarily women's careers and the influence of children on these career-paths. In the following sections of this thesis, the three essays present the backgrounds of the papers, hypotheses, theories and models for each paper, estimation strategies, estimation results, interpretation of findings and conclusions. The Thesis ends with various closing remarks.

1. What can Job Training do? Is University the Sole Way of Life?

- The Effects of School Education and Job Training on Wages in Korea -

Abstract

In contradistinction to Becker's human capital theory, more educated university graduates in Korea suffer from high unemployment. In spite of this fact, the desire to go on to university continues to grow and any job related education and training are neglected. This study aims to find the reason behind this bias in Korea concerning human capital investment. Employing the data taken from the 'Korean Labor and Income Panel Study (KLIPS),' the study estimates the effects of school education and job training on wages by means of the random effects and Heckman 2 step methods. The finding is that the effects of a university education on wages are much more profound than those of lower educational levels. The effect of job training turns out to be marginal compared with the effect of school education. This tends to lead Koreans to invest so much in school and university education whilst holding job training in low esteem.

Keywords: School education, Job training, Human capital, Wages

1.1. Introduction

In 2008, 83.8% of all high school graduates went on to university or junior college in Korea. This number is more than five times as high as it was in 1980. However, only 76.7% of graduates from all tertiary educational institutions and merely 69% of general university graduates found a job in 2008. Only 56.1% of these graduates have a regular job (Korean Educational Development Institute (2009)). In spite of the fact that university graduates are oversupplied in the labor market, the desire for a university degree continues to grow. On the other hand, the number of vocational high school graduates continues to fall. It is, in general, rather rare in Korea that people have any form of job training - either an apprenticeship or further training - after completing their school education: Only about 5% of all adults participate in any kind of job training. Why then does Korea invest so much in school/university education while neglecting job related education and training unusually?

A positive correlation between earnings and the level of skills – school education and on-the-job training – and the reduction of the unemployment rate with the increasing level of skills were introduced as the stylized fact in human capital theory by Becker (1962, 1964). However: in contradistinction to this fact, a high level of education, such as a university degree, is not assurance of a job in Korea. Mincer (1962), in parallel with Becker's development of human capital theory, studied on-the-job training investment as distinguished from more formal education. He proved the great importance of on-the-job training in comparison with formal education in the US economy. The economic effect of on-the-job training, which is virtually neglected in Korea, is critical to the theory of human capital investment. With these contradictions, one could raise the question as to the reasons behind this phenomenon and how this can be explicated by the use of Becker's theory.

Regarding this question, I would suggest some possible answers as follows: Korea gives a great deal of weight to school and university education, because the effect of graduating university on wages is much greater when compared with that of other lower educational levels. The wage difference between educational levels, however, can not be compensated for with any other human capital investment in Korea. Job training, which generally enhances one's productivity and earnings parallel with formal education, has only a marginal influence on wages. Koreans do not recognize job training as an alternative factor of human capital investment. This is the primary reason why people invest so much in school and university education and consider job training to be of no account.

Research into human capital investment in Korea has been revitalized since the late 1990's, when the Asian crisis severely hit the Korean economy: The unemployment rate, which had remained under 3% since 1988, increased greatly at the end of 1997.[3] In 1998, the unemployment rate grew to 6.95%. Job training became for the first time ever an important political measure to combat the high unemployment rate. Even at that time, empirical research on human capital investment becomes more important for the purpose of estimating the effects of job training on the economy. Reviewing research concerning human capital in-

3 In the month of December 1997 alone, approximately 3,000 companies went bankrupt, which led to a massive increase in the number of unemployed from 500,000 to 1,300,000.

vestment in Korea, Lyu (1995) first attempted to determine the economic effects of job training. He analyzed the data contained in the "Financial Statement Analysis" merging it with the wage data from the Ministry of Labor. Despite this pioneering effort, he could not find a single significant result concerning the effect of education and job training on wages. Kang and Noh (2000) estimated the effect of job training on employment and wages, using the 1998 data from the 'Korean Labor and Income Panel Study (KLIPS).' Employing the Heckman 2 step method in their analysis, they found that job training resulted in an increase in wages of approximately 20% while the effect of further education on wages, according to their analysis, was unexpectedly negative. They interpreted this negative impact of school education on wages as an anomaly and attributed it to the economic crisis of 1997 and 1998. Kim (2001) studied the productivity and efficiency of job training and its wage effect. In pursuit of this analysis, he convened the 'Skill Formation Survey' and the 'IT Industry Manpower Survey' of the Korea Labor Institute as a panel. Employing Almon's polynomial lagged model and the stochastic frontier model, he found a significant effect due to education and job training on a firm's productivity and efficiency. Following his analysis, however, the individual wage effect was determined to be insignificant. Kim (2002) further explored the individual wage effect of job training using data supplied by the 1998 and 1999 KLIPS panel. The job training effect on individual wages was found to be about 5% utilizing the first difference method, but the result was statistically insignificant at the 10% level.

Further studies of job training effects on wages proceeded in the field of 'skill biased technical change.' Kim and Kim (2007) and Kim (2009) analyzed the effect of job training on wage inequality. They concluded that job training has an influence on the growth of wage inequality. In the research, Kim and Kim (2007) found a 4-6% significant wage premium relating to job training using the 1998-2005 KLIPS panel data. Kim's research (2009) revealed a 6.6% significant wage premium attributable to job training by means of the switching regression model, using the 2005 and 2007 data from the Human Capital Corporate Panel (HCCP). In the field of 'skill biased technical change,' the earnings/wage inequality problem relating to education has been plentifully researched since the end of the 1980's, and the educational effect on wages was also indirectly discussed in these researches.[4] However, attention has seldom been paid to the education and wages in a human capital framework. Recently, Lee and Lee (2006) have analyzed the wage determinant in the Korean labor market using the

4 See Yoo (1989); Kwark and Rhee (1993); Ryoo (1993); Nam (1996); Choi, Jeong, and Jung (2005); and Kim (2005).

'Korean Labor and Income Panel Study (KLIPS)' by means of quantile regression. They found that age and education play important roles in the determination of wages in Korea. Kwack, Lee and Choi (2007) studied gender earnings gaps and return to schooling using the Mincer earnings function. They used data from the 'Survey Report on Wage Structure' (1993-2002) supplied by the Ministry of Labor. By means of OLS estimation, male and female workers with university and higher education earn 1.8 and 2.9 times more respectively than those with a middle school or lower education in 2002. In these studies, however, the estimation problems of the Mincer earnings function using the OLS method - measurement error, omitted ability bias and selection bias - are not generally considered. As a possible countermeasure to these problems, this study analyzes the social dependency of these biases. From this, the study deduces some plausible facts and uses them as guidance to find more consistent estimates.

The purpose of the study is to find the economic reason for the biased human capital investment behavior of Koreans concerning school education and job training. To test my hypothesis, this study estimates different wage effects between educational levels and job training. Concerning educational levels, the estimation focuses on the wage difference between senior high school and university graduation. To this end, I compare the effect of university education with that of job training, which is a new project in Korea.

The plan of the study is as follows: Section 1.2 introduces the education system in Korea. Section 1.3 outlines Becker's human capital theory and the Mincer earnings function, and presents the estimation problems of his earnings function by OLS. Considering the Korean educational system and other social factors, I assume certain plausible facts, related to the estimation problems. Section 1.4 introduces the data. This study employs the 'Korean Labor and Income Panel Study'. Descriptive statistics are shown in this Section. In Section 1.5, the estimation methods are outlined and the estimation results are interpreted. The random effects method and Heckman 2 step method are used in this study. Section 1.6 summarizes the estimation result and concludes this study.

1.2. Education System in Korea

The present school system in Korea, which is a single ladder system influenced by the USA, was established in 1949 according to the Law of Education for the Republic of Korea, founded in 1948. The school system in Korea mainly consists of 6 years elementary school, 3 years junior high school, and 3 years senior high school prior to taking either 2-3 years junior college or 4 years bachelor's course. It proceeds then with a 2 years master's course and a further 3 years and more for the doctoral course (see Table 1.1).

The Korean elementary school is equivalent to the *Grundschule* in Germany, which is classified as the primary level of education defined as 'International Standard Classification of Education-97' level 1 (ISCED-97 level 1). Three years junior high school, which is approximately identical to *Hauptschule, Realschule* or *Gymnasium* in Germany, is equivalent to the lower secondary level of education, ISCED-97 level 2. Senior high school is the upper secondary level of education classified as ISCED-97 level 3. This level of education is divided again into academic high school (ISCED-97 level 3A) and vocational - technical or business - high school (ISCED-97 level 3C). The academic high school is comparable to the *Gymnasiumoberstufe* or *Fachgymnasium* in Germany and vocational high school to the *Berufschule, Berufsfachschule, Berufsoberschule* or *Fachoberschule*. After completing academic high school, most pupils enter either 2-3 years junior college or 4 years university. Junior college is classified as ISCED-97 level 5B, which is the first stage of tertiary education with practical technical and occupational specifics in a short curriculum. Classified as ISCED-97 level 5A (1st), bachelor's courses at the university constitute the first stage of tertiary education with a strong theoretical foundation in a medium level curriculum. Some faculties, for instance, the medical faculty, have a curriculum of 6 years at this level of education. A postgraduate education consists of a 2-year master's course (ISCED-97 level 5A (2nd)), which is the first stage of an extended tertiary education, and a further 3-year and more doctoral course. This level provides sufficient qualifications for entry into advanced research programs and professions with high skills requirements. The doctoral degree constitutes the second stage of tertiary education (ISCED-97 level 6).

Kindergarten education is not acknowledged as an official school education in Korea yet. Education at the elementary school and the junior high school are compulsory. Further school education requires pupils to pass the appropriate entrance exam and to pay school fees. After graduating an academic senior high school, pupils go on to university or junior college. But graduates from a vocational senior high school can also enter either junior college or university if they pass the entrance examination. About 67% of the vocational school graduates go on to junior college. The remaining pupils at the vocational high school take 1-6 month practical vocational training in a company and voluntarily undergo an exam to obtain a vocational qualification in their 12th school year in general. Yet this practical training with little importance is merely a part of the curriculum of vocational high schools. After finishing school education, job training is generally carried out by the company, the government or on one's own initiative.

Table 1.1: School System in Korea

<table>
<tr><th>Age</th><th>Class</th><th colspan="2">Korean school system</th><th colspan="2">ISCED-97 level</th></tr>
<tr><td>24+
(Male: 26+)</td><td>19-21</td><td>Doctor's course</td><td rowspan="3"></td><td>6</td><td rowspan="3"></td></tr>
<tr><td>22-23
(Male: 24 to 25)</td><td>17-18</td><td>Master's course</td><td>5A (2nd)</td></tr>
<tr><td>21 (Male: 23)
20 (Male: 22)</td><td>16
15</td><td rowspan="2">Bachelor's course</td><td rowspan="2">5A (1st)</td></tr>
<tr><td>19 (Male: 21)
18 (Male: 20)[5]</td><td>14
13</td><td>Junior college</td><td>5B</td></tr>
<tr><td>15-17</td><td>10-12</td><td>Academic Senior High school</td><td>Vocational Senior high school</td><td>3A</td><td>3C</td></tr>
<tr><td>12-14</td><td>7-9</td><td colspan="2">Junior high school</td><td colspan="2">2</td></tr>
<tr><td>6-11</td><td>1-6</td><td colspan="2">Elementary school</td><td colspan="2">1</td></tr>
<tr><td>3-5</td><td>0</td><td colspan="2">Kindergarten</td><td colspan="2">0</td></tr>
</table>

Sources: *1) OECD (1999), 2) Kim, A.C. et al. (2006)*

Excessive competition for university entrance has become one of the most pressing social problems in recent years in Korea. This leads young teenagers to prepare for the entrance exam in advance in order to increase the probability of their acceptance by a (good) university. Most households are economically and mentally burdened with the problem of the education of their children. They sacrifice their own living expenditure and cultural activities in order to enable their children to have the chance of a good education. In 2005, supplementary educational expenses for the average Korean family reached 2.79% of GDP, with the result that private lessons for teenagers has been a rapidly growing area of the underground economy in Korea. This economic sector absorbs unemployed university graduates, thereby forming a big black market. Such private education imposes a tremendous burden on the family and discriminates against poor chil-

5 In Korea, male students between the ages of 18 and 30 are obliged to undertake about 2 years of compulsory military service.

dren. The usefulness of formal educations is undermined and this leads to the uniformity of secondary education in which pupils only prepare entrance exam. One doubts the efficiency of this educational fever and fierce competition for university entrance. Paul Krugman (1994) pointed out that increasing input growth with the upgrading of education and a powerful mobilization of economic resources in East Asia has not been accompanied by a corresponding growth in efficiency.

In spite of all such criticisms, however, these educational problems in Korea have been neither resolved nor lessened. Since 2004, more than 80% of all high school graduates have gone on to university or junior college. On the other hand, the participation rate in non-formal job training reached only 4.67% in 2002 in Korea (Lee et al. (2006), p. 38) as compared with 18% on average in the OECD in the same period (OECD (2007), p. 354). Experts said that the structure and quality of labor supply is inadequately adjusted demand for labor supply (Woo and Kim (2004)). Confronted with these problems, some educational experts are concerned about the impact of the single ladder school system. They have discussed whether this single ladder system has overstressed the value of a university education and caused an overinvestment in school education in general. Furthermore, they question whether this system impacts on the minimal interest in vocational education and job training (Korean Educational Development Institute (2006)). The German educational system is well-known in Korea with its long tradition of job training together with an apprentice training system (i.e. *Duales System*) at school. Respecting these concerns, the German educational system has begun to be seriously studied in Korea.

1.3. Theory, Model and Problems of Estimation

Becker's Theory and Mincer's Model

Becker (1962, 1964) formulates the general theory of human capital, on which this study is based. He illustrates the dependency of earnings on various factors of human capital such as on-the-job training, school education, other knowledge and information, emotional and physical health, and ability. According to the theory, earnings are the gross return on human capital investment, so there is a positive correlation between the amount of investment in human capital and the amount of earnings. School education, on-the-job training and other human capital investments enhance productivity, and positively influence the future earnings of a worker. In an individual's lifetime, human capital investments reduce earnings at younger ages because costs are deducted from earnings, but raise earnings at older ages because the returns of these investments are added to earnings then. As regards ability, Becker writes, an abler person tends to invest in human capital more than others, the distribution of earnings could be very unequal and skewed, although ability may be equally distributed. With respect to on-the-job training, any increased profitability through investment in on-the-job training is shared between an employer and an employee, depending on who finances the investments and who collects their returns.

Mincer (1974) develops in his seminal work "Schooling, Experience and Earnings" his human capital earnings function in the following way to explain the differences in earnings of people who differ in terms of schooling and ages:

$$\ln y = \beta_0 + \beta_1 s + \beta_2 EX + \beta_3 EX^2 + u$$

The dependent variable ln y is the logarithmic earnings of the individual under consideration. The independent variable s is a measure of their schooling whilst EX is equal to their labor market experience, which leads earnings to vary along with age. Schooling determines one's skill level, which consists of school education and on-the-job training, following Becker (1962). Labor market experience on the job is measured by a proxy variable which is equal to an individual's current age minus their years of schooling minus their age at the beginning of schooling, if actual experience cannot be quantified. Experience and its squared form determine a concave age-earnings profile varying with one's skill. This proxy variable for experience is, however, only appropriate for males, but not for females, whose labor market experience is interrupted due to bringing up children. To complement this proxy variable for labor market experience, studies additionally include tenure and its squared form of present employment. u is a random disturbance term. It also contains the unobserved determinants of earnings, such as ability. The coefficients β show the percentage changes in earnings.

Although the wage effect of human capital provides an appropriate explanation of the age-earnings profile, individual earnings can not be solely explained by human capital variables. Wages are differentiated by other socio-economic factors, for instance, some native social status attributed to individuals such as gender and race; some family characteristics such as marital status and responsibility for family – household head and having children; some firm characteristics for an employee such as industry, firm size and working contract, and some institutional factors such as minimum wages and trade unions.[6] Even for indi-

6 See Kalleberg and Sorensen (1979), Kalleberg (1988), DiPrete and McManus (1996), and Hollister (2004).

viduals with the same schooling level, their heterogeneity concerning socio-economic characteristics could lead to differences in earnings. Researchers include these variables to take account of such heterogeneity and their concomitant earnings effect. The model in this study, based on the Mincer earnings function, attempts to explain the wages of individuals utilizing variables of human capital such as school education level, job training, labor market experience and tenure controlling for other individual, family and work place characteristics.

In numerous studies at different times and in different countries, Mincer's human capital earnings function has successfully explained the 'age-earnings profile,' which is a widely accepted empirical specification in economics. In the practical application of Mincer's model however, the exact estimation of human capital effects on earnings is not totally unproblematic. His function suffers from the following problems.

Problems of Mincer Earnings Function and Plausible Assumptions for Korean Data

Measurement error, omitted ability bias and selection bias have been noted as difficulties of Mincer earnings function in the evaluation of the consistent schooling effect on wages by the OLS. As countermeasures, various solutions have been proposed, for example: Using a proxy variable or an instrumental variable, employing the fixed effects method, the Heckman 2 step method or the control function method etc. However, their success mainly depends on the availability of an appropriate dataset or the necessary variables. Still, researchers who want to obtain accurate estimations for the possible returns on human capital investment continue to meet with these problems. On the other hand, the gravity of these problems – measurement error bias, omitted ability bias and selection bias – depends on the systems and laws of education, social standards and cultural assumptions in the countries under consideration, and varies according to the particular social environment. So, an understanding of these dependencies could offer some helpful guidance in a more precise estimation of these results, which are more or less biased because of limitations of data or estimation methods. I begin by assuming certain plausible facts, related to the estimation problems of the Mincer earnings function, connected to social factors in Korea and use these as a plausible basis for this study.

A measurement error is generated by incorrect self-reporting about the level or duration of schooling or if the actual duration of schooling is not available. This type of error usually generates a downward bias in the OLS estimate of the schooling.[7] To remove measurement error, we could use an instrumental variable for schooling if we can find an appropriate instrument, which is often unsuccessful. The dimension of the bias caused by this kind of error could vary and be dependant on the system of education and social values in different countries or cultural areas. In Korea as well as in East Asia generally, school education is a factor that not only has an influence on wages but can also confer cultural status and social power on the individual. It is regarded as an important matter concerning one's own dignity as well as that of the family. People who have a minimal educational level are often unwilling to reveal their true level of schooling, which could lead to an increase the dimension of measurement error. Such overstatements of educational attainments could be prevalent in the generation of people aged between 60 and 70 who underwent social and political difficulties in Korea – the Korean War and its consequential poverty – in their youth, and hence, many of them could not afford to attain a certain level of education. Considering such social conditions, the proportion of junior and senior high school graduates, considered as the common level of education reached by people in the generation between the ages of 60 and 70, might be slightly larger in this data than the proportion in the population as a whole. Unlike the education variable, the measurement error due to job training is usually attributable to losing sight of the fact of whether people have any job training or not. There is no general motivation to overstate one's job training in Korea. It does not lead to a downward bias on the wage earned.

Another measurement error which is generated due to unavailability of the actual duration of schooling has an intimate connection to the educational system prevalent in their countries. In the German education system, individuals are permitted a variable number of years to complete the same level of education. In Korea, however, the completion of a school education depends on years of schooling. The students are considered to have completed their school education after a given number of years. In the case of Korea, the completion of a specific degree and the years of schooling themselves do not lead to a problematic measurement error.

Omitted ability bias is a well-known problem in the estimation of the education effect on wages. Becker (1962, p. 48) already pointed out that an abler person tends to invest more in human capital than others. Choice of job training could also be correlated with an unobserved ability if an abler person with high earn-

7 See Griliches (1979), Ashenfelter and Krueger (1992), and Ashenfelter and Zimmerman (1993).

ings potential participates in job training. Usually, the unobserved ability in the disturbance term is positively correlated with schooling/job training and earnings in the OLS estimation. The omitted ability generally biases the return to schooling/job training upwards.[8] Countermeasures against this bias could be to use a proxy variable of ability or an instrumental variable of schooling. For example, a study using the data of monozygotic twins furnishes us with an interesting method to counter the bias of omitted ability (Miller et al. (1995)). The problem of this approach, however, lies in how to find an appropriate variable. In the panel dataset, the fixed effects model has been proposed as an alternative to deal with this omitted ability bias whilst assuming that the omitted ability variable is constant in the observed period of time.

Ashenfelter and Zimmerman (1993, 1997) estimate returns attributable to schooling with data on brothers and on fathers and sons from the National Longitudinal Survey. They compared differences in schooling between brothers, fathers and sons with the differences in their respective earnings. The sample was selected to ensure similar innate ability and family backgrounds, therefore, controlling unobserved family attributes in a straightforward fashion. A 25% upward bias was attributable to the impact of omitted family background and a 25% downward bias to the effect of measurement errors for siblings. The conclusion was drawn that the omitted ability and the oppositely directed measurement error offset their respective biases in a sample of the same size.

Selection problem can be sub-divided into a sample selection problem and the self-selection of schooling. Unlike the standard omitted variable, the problem of sample selection bias arises due to the dependent variable: earnings are not randomly missing in the dataset. As a countermeasure, Heckman (1979) developed his 2-step estimation. The self-selection bias of schooling is generated if the greater earnings of individuals with higher educational levels arise not because of this higher education, but because individuals with greater earnings capacity choose to acquire more schooling. It leads to an upward bias on earnings. Garen (1984) and Heckman and Robb (1985) developed the control function approach to correct self-selection bias.

In relation to ***sample selection***, a special kind of sample selection problem in Korea should be mentioned here. As stated earlier, university graduates in Korea suffer from a high unemployment rate. This group is overrepresented in the offi-

---

8 See Behrman et al. (1980), Blackburn and Neumark (1991), and Ashenfelter and Zimmerman (1993).

cial labor market. On the other hand, fierce competition for a university place in Korea has created a market for private lessons for high school students to prepare them for the entrance examination. This market is a magnet for students and unemployed university graduates, especially female graduates, by virtue of its flexible working conditions and (high) earnings with the possibility of evading the payment of taxes. Despite all these advantages, which would raise the reservation wage for some university graduates, this population officially has no wages as employees. As a consequence, values for wage data are incomplete, which may strengthen the sample selection bias.

Self-selection bias could occur, for example, if someone who has a good family background or a good network enabling them to secure a qualified job would choose higher schooling or job training appropriate to such a (highly) qualified job hence attaining high earnings. This self-selection process regarding education could usually arise from the choice of educational level and the choice of a particular route and faculty of education. This choice of educational level depends, however, not only on self-selection. It is also influenced by the education system, for example, the law on compulsory education and other social circumstances.

In Korea, compulsory education requiring 6-years of schooling at the elementary level was introduced in 1948. Between 1984 and 2004, compulsory education was extended to include 3-years of junior high school, so together 9 years of schooling. After these 9 years of compulsory education, 84.5% of junior high school graduates went to senior high school in 1980, 90.7% in 1985 and 95.7% in 1990. This proportion increased to 99.8% in 2006. During these time periods, the entrance rate to junior college and university also increased rapidly: In 1980 the entrance rate to junior college and university was 23.7%, rising to 51.4% in 1990, and reaching 68% in 2000. Since 2004, the entrance rate to junior college and university has remained steady at more than 80% (Korean Educational Development Institute (1980-2009)). The data reveals a rapid upward tendency in schooling. Since the end of the 1980's, the common educational level has risen from senior high school to junior college or university.

In connection with the self-selection problem, this kind of enhancement of educational level does not cause a self-selection bias on wages and further reduces the potential of any self-selection bias in the dataset. Having successfully finished their high school education, people choose either to find a job or to attend junior college for 2 years or university for 4 years. It depends on the grade of the entrance exam and the economic resources of a family, that is, their ability to pay the college/university tuition fees, which is quite distinct from a self-selection process. Choosing which route to take in an advanced education among pupils entering either an academic senior high school or a vocational senior high school in the period of secondary education depends also on the results of a final examination and one's school records. The self-selection process for higher schooling actually arises after the completion of a bachelor degree in Korea. Approximately 12.9% of bachelor graduates proceeded to a master's course and 1.4% of master graduates proceeded to a doctorial course in 2008. Related to the self-selection bias, one should pay attention to which kind of variable for schooling or job training is relevant. The estimates of the effects of a master's or doctorial degree and job training should be prudently interpreted in the Korean dataset in reference to the self-selection bias.

Table 1.2: Biases of Mincer Earning Estimation in Educational Level and Job Training

Education level/Job training	**Measurement error bias**	**Omitted ability bias**	**Self-Selection bias**
Compulsory level	no	no	no
Low level	no	weak	no
Common level	strong	middle	no
High level	no	strong	strong
Job training	no	noticeable	noticeable

Concluding this section, Table 1.2 presents some plausible assumptions related to the estimation problems of the Mincer earnings function for the Korean dataset. Considering the Korean case, we can easily confirm that there is generally no measurement error, omitted ability bias or self-selection bias at the compulsory level of education, when the effects of educational level on wages are estimated. The measurement error would be relatively strong in the common level of education. Some people with minimal education might exaggerate their achievements and claim to possess a common level of education when responding to questionnaires in Korea. The omitted ability bias generally becomes stronger when considering a higher level education, because an abler person tends to invest more in higher level education. Self-selection bias may be just a problem at this higher level of education. Koreans usually try hard to attain at least the common level of education without any selective consideration. The job training variable does not plausibly suffer from any systematic measurement error, while the omitted ability and the self-selection biases could noticeably influence the job training effect on wages. There is no motivation to overstate one's job training in Korea.

1.4. Data and Descriptive Statistics

To test the hypotheses of this study, the 'Korean Labor and Income Panel Study (KLIPS)' published by the Korea Labor Institute, is used. KLIPS is a longitudinal survey of the labor market and income activities of households and individuals who live in cities and urban areas. Being the first survey produced by a domestic panel on labor-related issues, it has served as a data source for the micro-economic analysis of labor market activities and transitions, thereby contributing to the development and evaluation of labor market policies. In designing and managing KLIPS, its role models were a set of successful longitudinal surveys, conducted in industrialized countries, such as NLS (National Longitudinal Surveys), NLSy (National Longitudinal Survey of Youth), and PSID (Panel Study of Income Dynamics) in the USA, SLID (Survey of Labour and Income Dynamics) in Canada, BHPS (British Household Panel Survey), and GSOEP (German Socio-Economic Panel), among many others. Questionnaires were sent out to 5,000 households, constituting 13,321 household members in 1998 in 7 big cities and urban areas in 8 provinces. Respondents comprised 4,012 employees, 2,415 self-employed individuals and 7,315 unemployed people or the economically inactive population. The panel took into account a variety of variables concerning social and economic backgrounds: Family, housing, income, consumption, education, job training, working contract, vocations, working conditions, job-seeking experience, mobility on the labor market, health, marital status, job contentment, etc. Income is divided into wages, earnings for the self employed, financial income, immovable income, transfer income and other types of income.

In 1998, when this survey was launched, Korea was faced with an economics crisis for the first time in its post-industrial history. Job training was a main government policy to tackle mass unemployment. The negative income shocks experienced by Korea between 1997 and 1998 increased the number of the unemployed who participated in job training programs in 1998 and 1999 more than 80 fold. This number of participants reached 163,000 in 1998 and further increased to 171,000 in 1999 (it was just 1,947 in 1997). At that time, job training was financed out of government funds as an exceptional measure against mass unemployment. When studying job training, the exceptional economic situation in the years around 2000 means that a comparable estimate of any job training effect on wages in other periods is skewed. On the other hand, variables in KLIPS often changed at the beginning of the survey, or some of them were excluded from the questionnaire. The required variables employed in this study, date only from 2002. For these reasons, this study uses the KLIPS dataset between 2002 and 2006. In order to estimate the effects of education and job training on wages, this study selected the data for male adults aged between 26 and 55. At the age of 26, male university graduates in Korea usually begin vocational life after completion of about 2 years compulsory military service. Early retirement is an option from the age of about 55. This option also reduces the measurement error attributable to the education level of the elderly in Korea. Women were excluded from the analysis. As previously mentioned, the proxy variable for labor market experience used in this study is not an appropriate proxy for females.

The main variables used in this study are wages of the main job, education and job training. The mean value of the real monthly net wage of the employees is approximately 1,848,955 Won (i.e. 1,471 US$, base year: 2000) in the dataset with 6,649 observations comprised of 2,545 people. The natural log of the wage demonstrates an approximately normal distribution with a mean value of 5.2 (see Figure 1.1). Standard deviation is about 0.48. Concerning education levels, approximately 3.42% of the individual observations had completed an elementary school education, 8.79% junior high school education, 40.35% senior high school education, 14.47% junior college education, 28.61% had received a bachelor's degree and 3.70% a master's degree. About 0.66% of them had a doctoral degree.

Concerning the job training variable, the KLIPS basically reports in each survey about job training in the last year under consideration. In the first year survey, the KLIPS further inquires about job training experience in one's earlier working life. As asserted by Becker (1962), the influence of educational level on wage could increase throughout an individual's entire working life. In order to estimate a comparable wage effect of job training with educational level, this study creates a job training variable combining each year's information concerning job training with the information of job training experience of one's earlier working life as reported in the first survey of the KLIPS: i.e. job training experience, whether individuals have ever completed any job training course in their whole working life. Among 6,649 observations, 30.46% had received job training experience in their whole working life, while 15.47% had completed some job training in the last year under consideration.

The average labor market experience is 16.6 years in the dataset. Experience is utilized by 10 dummy variables, which are approximately classified in a decile form in 10 experience groups in this study. The average tenure is 6.3 years. The tenure effect on wage is also measured by 10 dummy variables in the same way as the experience. This study additionally includes variables such as 5 year dummies, 16 dummies of residence of house, 202 industry dummies, 19 firm type dummies, 8 firm size dummies, 5 casual worker dummies, and dummy variables for regular worker, part time, overtime, union, shift work, household head and children. Additionally, this study employed 4 house ownership dummies, natural log household sustenance allowance for parents, natural log non household labor income, natural log household financial wealth and natural log household debt in order to correct the sample selection bias in the estimation of the Heckman 2 step.[9] The descriptive summary of the aforementioned variables is contained in the appendix (see Tables 1.3 and 1.4).

1.5. Estimation Methods and Results

For the estimation, the random effects and the Heckman 2 step method are used. According to the studies by Ashenfelter and Zimmerman (1993, 1997), as outlined in Section 1.3, the problems of the OLS estimation of Mincer earnings function concerning education variable could be reduced to the selection bias. Following plausible assumptions in this study, this selection bias concerning education may not be a problem in the common level of education. i.e. senior high school graduation to bachelor's degree in Korea, on which this study mainly focuses. Concerning biases attributable to job training variable, the fixed effects method could remove the unobserved time invariant ability bias and the control function method offers a possibility to remove the self-selection bias. However, utilizing the fixed effects method, this study encounters a problem: the education level for adults, who are between ages of 26 and 55, is usually time invariant.[10] This study only utilizes the fixed effects method as a reference estimator for estimating the job training effect and compares its result with that obtained by the random effects method. The control function, which is proposed by Heckman and Robb (1985) as a technique for removing the self-selection

---

9 For the strategies for utilizing variables, this study refers to Entorf and Kramarz (1997).

10 When considering the changing education level of adults, this wage effect can differ greatly from the usual education effect. Professor Roland Eisen critically pointed out in his lecture "Arbeitsökonomik" (2003) that education is not only an investment good. People often consume educational programs, which is often neglected in the studies of human capital investment. In many cases, participation in an educational course for an adult does not mean human capital investment in the sense of Becker's theory (1962, 1964).

bias attributable to job training, revealed weak points due to an insufficiency of correction variables and exceeding sensitive in estimates given by the correction term (Fitzenberger and Prey (1998), pp. 51). As a countermeasure, there are further proposals, such as pre-program test (LaLonde (1986)) and method of matching (Rosenbaum and Rubin (1983) and Lechner (1996a, b)), having different advantages but showing different limits, too. Such a confrontation with self-selection bias concerning job training will be studied as a separate subject in a future paper. The self-selection problem attributable to job training is not dealt with in this study.

The advantage of the random effects method is its competence to remove the serial correlation in the error term associated with individual heterogeneity employing panel data. Moreover, its transformation technique enables the utilization of time invariant variables such as adult educational level in the estimation unlike the fixed effects method. As stated previously, the increasing number of university graduates with a low employment rate could be related to sample selection bias. Concerning this, Heckman's theory offers a possible explanation. According to his theory, some people decide not to participate in the labor market, because the market wage is lower than their at least expected reservation wage. The Heckman 2 step method could expand the object of the study by correcting the sample selection bias. It enables one to examine not only employees but also indirectly the self-employed, the unemployed and the economically inactive population. Concerning the question of this study as to why Korea invests so much in school/university education whilst, simultaneously neglecting job related education and training, this broad view of analysis could support us in the search for a more general answer to the question.

1.5.1. Random Effects Estimation

The Random Effects Method

The random effects method removes the serial correlation in the error term through the GLS (generalized least square) transformation in the following way:

$$y_{it} - \lambda \bar{y}_i = \beta_0(1-\lambda) + \beta(x_{it} - \lambda \bar{x}_i) + v_{it} - \lambda \bar{v}_i,$$

where $\bar{y}_i = T^{-1}\sum_{t=1}^{T} y_{it}$, $\bar{x}_i = T^{-1}\sum_{t=1}^{T} x_{it}$, $\bar{v}_i = T^{-1}\sum_{t=1}^{T} v_{it}$,

$v_{it} = a_i + u_{it}$ and $\lambda = 1 - \left[\sigma_u^2 / (\sigma_u^2 + T\sigma_a^2)\right]^{1/2}$.

Where, y is a dependent variable, which is log monthly net wages in this study. x is independent variable, which is variables of educational levels and job training, and other control variables in this study. σ_a^2 is the variance of a_i, which is unobserved individual effect (i.e. the random effect). β_0 is a constant term and β coefficient. T represents the observed time. In this study, this is 5 years from 2002 to 2006. σ_u^2 is the variance of idiosyncratic error u_{it}, which changes across time t as well as across individuals i. a_i is assumed to be independent of the observed explanatory variables x, however, causes a serial correlation with u_{it} across time periods. λ shows us the size of the serial correlation. If λ =0, then the function is a pooled OLS. In this case, the unobserved effect a_i is not important. If λ =1, then the function is the same as the fixed effects model. Growing σ_a^2, λ becomes closer to 1.

Estimation Results

Table 1.5 presents the main estimation result of this study: The wage effect of educational levels and job training. The base group of educational variables is the dummy variable that concerns the completion of senior high school education. The base group of job training variable – i.e. job training experience which has been amassed in one's whole working life – is the dummy that deals with the absence of job training. The results in general show an enhancement of education effect on wages when increasing educational level and a positive effect due to job training on wages.

The second and third columns in this table show the estimation result obtained using the OLS. Estimates are statistically significant at the level of 1% with the heteroscedasticity-robust standard error. The model in the second column estimates the job training effect on wages using the variable of job training experience in one's whole working life, while the model in the third column, as a reference, uses the completed job training in the last year under consideration. A completed elementary school education leads to an approximately 16% lower effect on wages than the effect observed in the base group. In the case of a completed junior high school education, it is approximately -11%. Educational effect on wages is 8% higher for those who graduated from junior college, about 24% for a bachelor's degree, 55% for a master's degree and 71% for a doctorate compared to the base group. Concerning job training, the effect of job training experience is approximately 4.3% higher than the effect of no job training. This is lower than the wage effect of completed job training in the last year under consideration. i.e. 6.8%.

The sixth and seventh columns present, the estimation result arrived at by the random effects technique: The effects of education levels and job training on wages for 2,545 male employees. Estimates are statistically highly significant with the heteroscedasticity-robust standard error. A completed elementary school education has an approximately 15% lower wage effect than the effect observed in the base group. Completed junior high school education leads to a wage effect about 13% less. This effect is about 9.4% higher for those who graduated from junior college, 27.5 % for a bachelor's degree, 54% for a master's degree and 80% for a doctorate compared to the base group. Concerning job training, the effect job training experience has on wages is approximately 3% higher compared with the effect of having no job training. This is slightly lower than the estimate of job training obtained from the last year under consideration (3.6%). By means of the fixed effects, the job training experience demonstrates a 1.8% wage effect at about 15% significance level with the heteroscedasticity-robust standard error. Job training in the last year has a 2.3% significant wage effect. Removing time invariant ability bias, the influence of job training becomes weaker.

In reference to the hypothesis of this study, the estimation results demonstrate that university graduation leads to 27.5% higher wages for employees compared with senior high school graduation. Junior college graduation leads to a 9.4% wage increase compared to senior high school graduation. Master's degrees and doctoral degrees could lead to a 54% and an 80% wage enhancement respectively. However, employees who had participated in any job training programs in their whole working life could just expect 3% wage increases compared with those who have never participated in any job training. The estimates obtained by the random effects method offer numerical values supporting my hypothesis.

1.5.2. Heckman 2 Step Estimation

The Heckman 2 Step Method

Concerning the counterintuitive low employment rate witnessed in university graduates in Korea, this study suspects the sample selection bias in the wage data. To correct the sample selection bias, the Heckman 2 step method computes the causality and probability of a sample selection and transfers the missing data problem of dependent variable to an ordinary omitted explanatory variable problem using the inverse Mill's ratio. Gronau (1974) developed a labor supply model correcting the sample selection bias. Following this model, a person decides to participate in the labor market only if market wage is larger than reservation wage. This is that level of wage, which just motivates people to take a job in the labor market:

$$y_i^0 - y_i^r = x_{i1}\beta_1 + u_{i1} - x_{i2}\beta_2 - \gamma_2 a_i - u_{i2} > 0,$$

where $y_i^0 = x_{i1}\beta_1 + u_{i1}$ represents the market wage and $y_i^r = x_{i2}\beta_2 + \gamma_2 a_i + u_{i2}$ is the reservation wage. Herewith, x_{i1} is the vector of explanatory variables for the market wage of any person i. a_i is one's non-wage income and x_{i2} is the variable vector, which is influenced by one's reservation wage level. Let $y_1 = x_{i1}\beta_1 + u_{i1}$ and y_2 be the binary labor force participation indicator, then Gronau's model can be written

$$y_1 = x_{i1}\beta_1 + u_{i1},$$

$$y_2 = 1[x_i\delta_2 + v_{i2} > 0],$$

where $x_i\delta_2 + v_{i2} \equiv y_i^0 - y_i^r$ and x_i is the variable vector, which determines the labor market participation. In this model, the observations are concentrated around the zero wage point, because the wages of those who do not participate in the labor market are regarded as a zero wage. Thus, the selection function y_2 is incidentally truncated. This violates the Gauss-Markov assumption of zero correlation between independent variables and the error term. Ignoring index i, we find that

$$E(y_1 \mid x, y_2) = x_1\beta_1 + \gamma_1 E(v_2 \mid x, y_2) = x_1\beta_1 + \gamma_1 h(x, y_2),$$

where $h(x, y_2) = E(v_2 \mid x, y_2)$. Because the selected sample has only $y_2 = 1$, we need so $h(x,1)$. $h(x,1) = E(v_2 \mid v_2 > -x\delta_2) = \lambda(x\delta_2)$ is the inverse Mills ratio used by Heckman (1979). The inverse Mill's ratio is the ratio of the probability density function over the cumulative distribution function with the truncated normal distribution:

$$\lambda(x\delta_2) = \phi(x\delta_2 / \sigma_2) / \Phi(x\delta_2 / \sigma_2),$$

where σ_2 is the standard deviation of the error of y_2. What we estimate by means of the Heckman 2step approach is then the wage equation,

$$E(y_1 \mid x, y_2 = 1) = x_1\beta_1 + \gamma_1\lambda(x\delta_2),$$

as follows: First, estimate the probit function by means of maximum likelihood in order to acquire the estimates of δ_2. For each observation in the selected sample, calculate the inverse Mill's ratio. Estimate β_1 and $\gamma_1 = \rho\sigma_1$ by OLS regression y_1 on x_1, $\lambda(x\delta_2)$, where σ_1 is the standard deviation of the error of y_1 and ρ the correlation of σ_1 and σ_2. Following the Heckman 2 step estimation procedure, this study derives the selection equation in the following way: As the binary labor force participation indicator, we use the wage dummy in the selection equation. The explanatory variables are indicators of years, experience, education, job training, marital status and household head, and household characteristics such as dummy variables for children, type of house ownership, residence, sustenance allowance for the parents, non-labor income, financial wealth and debt. In the wage equation, the wages are explained by the same independent variables which have already been described.

Estimation Results

Table 1.5 further presents the estimation result obtained by means of the Heckman 2 step method. The number of observations comprises 1,318 and 549 observations among these are withheld. The coefficient of the inverse Mill's ratio ($\gamma_1 = \rho\sigma_1$) is highly significant and positively signed which suggests that the error terms in the selection and wage equations are positively correlated. The probit estimation of the selection equation presents the following marginal effects: 1. Job training experience has a highly significant influence on the labor market participation with a value of 0.147, while the education variables do not present significant influence. 2. The sustenance allowance for parents has a significant positive influence with a value of 0.029. 3. The household debt has a highly significant negative impact with a value of -0.096. 4. Uncertain dwelling which cannot be classified as the usual type of house ownership - such as own house, leasing a house - shows a significant negative correlation with participation in the labor market with a value of -0.187. 5. In reference to those individuals with less than 5 years market experience, an experience of 5-9 years positively influences on their participations in the labor market, but 14 years plus experience negatively influences on the labor market participation. 6. A household head gives an approximately 15.5% points positive effect on labor market participation. 7. Most residence dummies show relative strong impacts on labor market participation. Related to the base group, located in Seoul, other residences mostly demonstrate strong negative effects. 8. The non labor income of a household presents a meager influence on the participation with value -0.011 around 15% significance level (see Table 1.6).

Analyzing the marginal effect of education on wages by the Heckman 2 step, the estimate of elementary school graduation gives us a much lower value than the estimate obtained by the OLS. It is approximately -36%. i.e. if we consider the unemployed, self-employed, or economically inactive population in the analysis, wage difference between elementary school and senior high school graduates becomes much more marked than the effect remarked upon by means of the OLS (-16%). The wage effect of attending junior high school is insignificant using the Heckman 2 step method. The wage effect is 17.4% higher for junior college graduates. This effect is 30% higher for those with a bachelor's degree and 71% higher for those with a master's degree than the effect noted for the base group. The estimate of the wage effect of possessing a doctoral degree is about 89%. These results are statistically significant up to 5% level with the heteroscedasticity-robust standard error. Regarding job training: After controlling for the effects of job training on participation in the labor market, the wage effect of one's job training experience throughout one's working life becomes insignificant up to the 10% level, although the estimate results in almost the same value (3%) as the estimate arrived at by the random effects method.

In reference our fundamental hypothesis, if we expand the object of analysis to non-employees through correcting the sample-selection bias, the obtained values concerning wage difference between education levels become stronger, thus, more certainly supporting the hypothesis. Compared to senior high school graduation, graduation from junior college results in a 17.4% higher effect and bachelor's degree a 30% higher effect on wages. However, the effect of job training experience becomes less significant.

1.5.3. Other Findings

Experience and Tenure This study obtained the labor market experience effect on wages for employees using 10 dummy variables. Using the random effects method, experience in the labor market of 5-9 years constitutes approximately 10%, 10-13 years 15%, 14-18 years 21.5%, 19-23 years 22%, 24-28 years 24%, 29-33 years 20% and 34-40 years 12.5% higher wage effects compared to an experience of less than 5 years up to 5% significance level. The experience effect is most noticeable in the period between 24 and 28 years. By using the Heckman 2 step method, this study found the following significant marginal effects: Experience of 5-9 years has a 7.5% higher impact on wage than that of the base group at about 10% significance level. 10-13 years experience has a 26% impact, 14-18 years a 33% higher impact on wages. Between 19-23 years experience leads to an effect about 34.6% higher on wages. 24-28 years a 29% and

29-33 years a 28% significantly higher wage effect compared to an experience of less than 5 years. The usual measure of experience - experience and experience squared - shows high significant effects on wages by the random effects and the Heckman 2 step methods. Estimation results well support the typical concave experience-earnings profile by both methods. Using the random effects, at 29 years of experience, the experience-wage profile arrives at the maximum value, while 20 years of experience achieve a maximum value using Heckman's method. The Heckman 2 step method demonstrates the higher influence of experience on wages compared with the result obtained by random effects (see Table 1.7, Figures 1.2 and 1.3).

Tenure, when analyzed using the random effects method, has highly significant effects on wages. One year of tenure leads to a 3.5% higher wage effect, 2 years 5.4%, 3 years 6.7%, 4 years 8%, 5-6 years 10%, 7-9 years 15%, 10-13 years 21.6%, 14-21 years 31.5% and more than 21 years 38% higher wage effects compared with tenure shorter than one year. In the usual measure of tenure – tenure and tenure squared – estimate of tenure demonstrates with a value of 0.018 highly significant wage effects, while estimate of tenure squared is insignificant. Using the Heckman 2 step method, only tenure higher than 9 years results in significant effects: 17.2% for 10-13 years, 13.7% for 14-21 years and 53.6% for more than 21 years of tenure. In the usual measure of tenure, estimate of tenure is insignificant and estimate of tenure squared shows a significant and slightly progressive trend equal to 0.00071 (see Table 1.8, Figures 1.4 and 1.5).

Family When the attention is focused on family background, the estimates given by a random effects analysis demonstrate a 10% and by a Heckman 2 step analysis a 20% significantly higher wage effect for married couples compared with singles. Further, a separated male significantly earns 46% more using Heckman's method and a divorced male 7% more using random effects method. The much higher values attributed to separated individuals in comparison with married couples are implausible. It might contain a measurement error concerning the wage: Separated men might overstate their wages. The wage of household head is 9.5% significantly higher than other household member and employees with children have approximately 3% higher wage than those without children according to the random effects method. In regard to place of residence, most regions of Korea have significantly negative effects on wages in comparison with Seoul (i.e. the capital city) using both estimation methods (see Table 1.9 and Table 1.12).

Firm When we consider the working contract and working conditions, regular labor has a significantly positive influence on wages as opposed to irregular labor. i.e. 9% by the random effects and 17.4% by the Heckman 2 step. Casual labor presents -8% lower influences using the random effects method in reference to non-temporary labor. Day labor demonstrates a 20% higher influence according to Heckman 2 step method in reference to non-temporary labor, while this is -8.5% by the random effects. Within employees, day laborers may comprise the low wage sector. If correcting the sample selection, however, hard working by a day laborer seems to increase wages compared to non-temporary labor. No overtime has a -3% and working shift a -4% lower wage effect by the random effects compared to working overtime and non-shift work up to 5% significance level respectively. The presence of a trade union has an 11% (Heckman 2 step) and a 4% (random effects) higher wage effect at the 1% significance level. The size of a firm has a highly significant influence on wages in all categories: The larger the firm, the greater the effect. As regards the type of firm, the Heckman 2 step method presents a -9.7% statistically significant wage effect concerning firms with government investment or public businesses compared to private companies. The random effects demonstrates that, among other results, the private educational institutions show 38.4% higher wage effects compare to private companies (see Tables 1.10, 1.11 and 1.13).

1.6. Conclusion

A primary question considered in this study is why people invest so much in school/university education in Korea, while job training is neglected. In order to find the answer, this study initially estimated the effect of school education and job training on wages employing the random effects and the Heckman 2 step methods. Summarizing the estimation results, we could note the following points:

Education Using the random effects method, this study found that employees who have completed elementary school earn approximately 15% less in wages than employees who have completed senior high school. The wage level of junior high school graduates is 13% lower than the wage level of senior high school graduates. Junior college graduates earn about 9.4% more. The wage level of graduates with a bachelor is approximately 27.5%, a master graduates 54% and doctors 80% higher than that of senior high school graduates. By the Heckman 2 step method, the wage difference between educational levels becomes larger: wage level of elementary school graduates is approximately 36% lower than that of senior high school graduates. This level of junior college graduates is about

17.4%, graduates with a bachelor 30% and master graduates about 71% higher than that of senior high school graduates. A doctoral degree holder could acquire about 89% higher wage compared to senior high school graduates.

Although the methods employed in this study could not technically remove the self-selection bias attributable to educational level, estimates for bachelor degree and lower may not be problematically biased due to this selection bias according to plausible assumptions made in this study. In contrast to the possession of a master's degree or a doctorate, graduation from senior high school to bachelor's degrees are considered to be a common level of education for the general Korean populace, which people should seek to attain without an undue consideration of special financial reward. This lessens the self-selection bias of these education groups. Following on from this, the estimation results, the 27% to 30% higher wage effect of a bachelor's degree as compared to the effect of a senior high school graduation, supports the hypothesis of this study. This noticeable wage difference between university graduates and senior high school graduates reasonably attracts people intending to go on to university.

Job Training By analyzing job training experience undertaken throughout one's whole working life, this study acquires an estimate = 3% by means of the random effects estimation. Comparing completed job training in the last year under consideration, the wage effect of job training experience is slightly low. The reason may be that the effect of job training on wages does not hold in the long run in Korea.[11] Removing the time invariant ability bias, the wage effect of job training becomes smaller. By means of the fixed effects method, the experience of job training leads only to a 1.8% wage increase (around 15% significance level). Expanding the object of our analysis to non-employees by the Heckman 2 step method, the wage effect of job training is no more significant although job training demonstrates a significant marginal effect on one's participation in the labor market. In reference to the hypothesis, the wages of relatively small group of employees who participated in any job training in their working life[12] are just 3% higher compared to the wages of those who have never taken any job training. Including non-employees in the analysis, job training does not significantly explain this effect concerning wage level.

Other Findings Dummy variables of experience and experience-wage profile show significant positive impacts on wages. This reflects the high dependency

---

11 Refer to the estimation results in Table 2.15.

12 As referred, about 5% of all adults participate in any job training course per year.

of wages on age in Korea. Tenure seems to increase linearly one's wage along with its tenure-wage profile. However, tenure averages only 6.3 years for male adults between the ages of 26 and 55. This short duration of tenure will strongly limit the tenure effect on wages. Concerning the working contract and working conditions, precarious working contracts – i.e. irregular, temporary and part time labor – lead to an 8% to 17% wage loss compared to a secured working contract. The existence of a trade union in the firm leads to a wage enhancement and the size of a firm has a positive influence on wages. So, employees who work in a big company with an organized trade union enjoy relatively high wages and better working conditions, which describe well the Korean industrial relationship. Related to family background, married couples, people with children and heads of households earn a significantly higher wage as compared to singles, people without children and non-household heads. So, the wage distribution benefits the traditional family. Regarding residence, the capital city, Seoul, has the highest wage level compared with other cities and urban areas. Employees in Seoul, which indisputably offers the best chance of education and getting a qualified job, earn approximately up to 30% more than people in other cities and urban areas. According to human capital theory, we could thereby find a reason why one-fourth of the population in Korea lives in Seoul and its peripheries, which is recognized as a serious social problem.

As we have outlined, there are significant wage differences between people who quit education after completing senior high school and those who proceed to a higher level of education. 2 years of junior college education increases one's wage by 9%-17%, and 4 years of bachelor education enhances the wage by 27% to 30%. Although job training is directly oriented to an enhancement of productivity, the wage with job training just marginally differs from the wage without any job training. Job training experience in one's whole working life increases the wage only up to 3%. From this, job training could hardly be recognized as a useful way to enhance one's wage. In 2007, 'Korean Labor and Income Panel Study (KLIPS)' questioned people as to whether they wanted any job training, and if they didn't want it, what the reason was. The answer shows that only 3.14% among 1,431 male respondents aged between 26 and 55 wanted to participate in any job training and 96.86% did not want it. Concerning the reason, 89.71% of respondents answered, they did not see any necessity for job training. In this situation, the necessary school education required to go on to university is regarded as a unique way to enhance their wages, when considering human capital investment. Undoubtedly, we know that there are other social factors, which attract people to university. Setting aside this fact, the economic meaning of job training is unnoticeable for the Koreans, which leads to a great deal of investment in school/university education, neglecting job training as was predicted in our hypothesis. A policy, which enhances productivity and returns of job training, would be desirable in Korea.

Beyond seeking to prove the hypothesis, I raise a new question: The effect of education and job training on wages in Germany, where more weight is given to job related education and vocational training. Further, a detailed consideration of the self-selection problematic associated with job training would be an interesting subject for the further study of job training and wages in Korea.

References

Ashenfelter, O. and Krueger, A. (1992), "Estimates of the Economic Return to Schooling from a New Sample of Twins," NBER Working Papers, 4143, National Bureau of Economic Research.

Ashenfelter, O. and Zimmerman, D.J. (1993), "Estimates of the Returns to Schooling from Sibling Data: Fathers, Sons, And Brothers," NBER Working Papers, 4491, National Bureau of Economic Research.

Ashenfelter, O. and Zimmerman, D.J. (1997), "Estimates of the Returns to Schooling from Sibling Data: Fathers, Sons, And Brothers," *Review of Economics and Statistics*, 79(1), pp. 1-9.

Becker, G.S. (1962), "Investment in Human Capital: A Theoretical Analysis," *Journal of Political Economy*, LXX, pp. 9-49.

Becker, G.S. (1964), *Human Capital*, New York: Columbia University Press.

Behrman, J.R., Hrubec, Z., Taubman P. and Wales, T.J. (1980), *Socioeconomic Success: A Study of the Effects of Genetic Endowments. Family Environment and Schooling*, Amsterdam: North-Holland.

Blackburn, M.L. and Neumark, D. (1991), "Omitted-Ability Bias and the Increase in the Return to Schooling," NBER Working Papers, 3693, National Bureau of Economic Research.

Choi, K.S., Jeong, J.H. and Jung, J.H. (2005), "The Rising Supply of College Graduates and Declining Returns for Young Cohort: The Case of Korea," *Global Economic Review*, 34(2), pp. 167-180.

DiPrete, T.A. and McManus P.A. (1996), "Institutions, Technical Change, and Diverging Life Chances: Earnings Mobility in the United States and Germany," *American Journal of Sociology*, 102(1), pp. 34-79.

Entorf, H. and Kramarz, F. (1997), "Does Unmeasured Ability Explain the Higher Wages of New Technology Workers?" *European Economic Review*, 41, pp. 1489-1510.

Fitzenberger, B. and Prey, H. (1998), "Beschäftigungs- und Verdienstwirkungen von Weiterbildungsmaßnahmen im ostdeutschen Transformationsprozeß: Eine Methodenkritik," in: Pfeiffer, F. and Pohlmeier, W. (Eds.), *Qualifikation, Weiterbildung und Arbeitsmarkterfolg*, ZEW Wirtschaftsanalysen, 31, Baden-Baden: Nomos, pp. 39–95.

Garen, J. (1984), "The Returns to Schooling: A Selectivity Bias Approach with a Continuous Choice Variable," *Econometrica*, 52(5), pp. 1199-1218.

Griliches, Z. (1979), "Sibling Models and Data in Economics: Beginnings of a Survey," *Journal of Political Economy*, 87(5), pp. 37-64.

Gronau, R. (1974), "Wage Comparisons: A Selectivity Bias," *Journal of Political Economy*, 82, pp. 1119-1149.

Heckman, J.J. (1979), "Sample Selection Bias as a Specification Error," *Econometrica*, 47(1), pp. 153-161.

Heckman, J.J. and Robb, R. (1985), "Alternative Methods for Evaluating the Impact of Interventions," in: Heckman, J.J. and Singer, B. (Eds.), *Longitudinal Analysis of Labor Market Data*, Econometric Society Monographs Series, 10, Cambridge: Cambridge University Press, pp. 156-245.

Hollister, M.N. (2004), "Does Firm Size Matter Anymore? The New Economy and Firm Size Wage Effects," *American Sociological Review*, 69, pp. 659-676.

Kalleberg, A.L. (1988), "Comparative Perspectives on Work Structures and Inequality," *Annual Review of Sociology*, 14, pp. 203-225.

Kalleberg, A.L. and Sorensen, A.B. (1979), "The Sociology of Labor Markets," *Annual Review of Sociology*, 5, pp. 351-379.

Kang, S.H. and Noh, H.S. (2000), "The Effect of Job Training on Employment and Wages in Korea," *Korean Journal of Labor Economics*, 23, pp. 127-151 (in Korean).

Kim, A.C., Park, J.Y., Kim, H.J., Yoon, J.H. and Kim, T.E. (2006), *The Reorganization Program of the School System for the Future*, Korean Educational Development Institute (in Korean).

Kim, A.K. (2001), *The Economic Gain and Distribution of Job Training in the Firm*, Dissertation, University of Korea (in Korean).

Kim, A.K. (2002), "A Study on the Individual Wage Effect of Job Training," *Korean Journal of Labor Economics*, 25, pp. 131-160 (in Korean).

Kim, A.K. (2009), "Wage Inequality and Job Training," Working Paper, 2009 Conference of Economics, Korean Labor Economic Association (in Korean).

Kim, C.H. and Kim, H.S. (2007), "The Effect of Job Training on Wage Inequality," Working Paper, the 8th KLIPS (Korean Labor and Income Panel Study) Conference, Korea Labor Institute (in Korean).

Kim, J.S., Lee, B.H. and Park, S.J. (2004), *An Evaluation of the Efficiency of Vocational Education and Training*, Policy Research, 2004-13, Korea Labor Institute (in Korean).

Korean Educational Development Institute (1980-2009), *Education Statistics*, Seoul, http://cesi.kedi.re.kr/index.jsp, (2010.06.09) (in Korean).

Korean Educational Development Institute (2009), *2008 Employment Statistics*, http://cesi.kedi.re.kr/index.jsp, (2009. 10.15), Seoul (in Korean).

Krugman, P. (1994), "The Myth of Asia's Miracle," *Foreign Affairs*, 73(6), pp. 62-78.

Kwark, N.S. and Rhee, C. (1993), "Educational Wage Differentials in Korea," *Seoul Journal of Economics*, 6(1), pp. 1-35.

Kwack, S.Y., Lee, Y.S. and Choi, H.S. (2007), "Gender Earning Gaps and Return to Schooling in Korea," *Journal of Economic Research*, 12, pp. 79–101.

LaLonde, R.J. (1986), "Evaluating the Econometric Evaluations of Training Programs with Experimental Data," *American Economic Review*, 76(4), pp. 604-620.

Lechner, M. (1996a), "An Evaluation of Public Sector Sponsored Continuous Vocational Training Programs in East Germany," IVS Discussion Paper Series, 539, Institut für Volkswirtschaft und Statistik (IVS), University Mannheim.

Lechner, M. (1996b), "The effects of Enterprise-related Continuous Vocational Training Programs in East Germany in Individual Employment and Earnings," IVS Discussion Paper Series, 542, Institut für Volkswirtschaft und Statistik (IVS), University Mannheim.

Lee, B.H., Kim, J.S., Kim, T.J., Jang, S.M. and Jang, W.S. (2006), *A Study of Life Long Education and its Effects on the Labor Market*, Policy Research, 2006-08, Seoul: Korea Labor Institute (in Korean).

Lee, B.J. and Lee, M.J. (2006), "Quantile Regression Analysis of Wage Determinants in the Korean Labor Market," *Journal of the Korean Economy*, 7(1), pp. 1-31.

Lyu, J.S. (1995), *Determinants and Effects of Investment in Job training on the Firm in Korea*, Research, 9510, Labor Administration and Training Institute (in Korean).

Miller, P., Mulvey, C. and Martin, N. (1995), "What do Twins Studies Reveal about the Economic Returns to Education? A Comparison of Australian and U.S. Findings," *American Economic Review*, 85(3), pp. 586-599.

Mincer, J. (1962), "On-the-Job Training: Costs, Returns, and some Implications," *Journal of Political Economy*, LXX, pp. 50-79.

Mincer, J. (I974), *Schooling, Experience, and Earnings*, New York: Columbia University Press.

Nam, Y.S. (1996), "Schooling and Changes in Earnings Differentials by Gender in South Korea, 1976-1991," *Economics of Education Review*, 15(3), pp. 245-258.

OECD (1999), *Classifying Educational Programmes – Manual for ISCED-97 Implementation in OECD Countries*, Paris.

OECD (2007), *Education at a Glance OECD 2007 Indicators*, Paris.

Rosenbaum, P.R. and Rubin, D.B. (1983), "The Central Role of the Propensity Score in Observational Studies for Causal Effects," *Biometrika*, 70(1), pp. 41-55.

Ryoo, J.K. (1993), "Changing Rates of Return to Education over Time: A Korean Case Study," *Economics of Education Review*, 12(1), pp. 71-80.

Woo, Ch.S. and Kim, H.M. (2004), *A Study of the Optimal Number of Individuals in a Highly Educated Labor Force*, Policy Research, 2004-4/5, Seoul: Ministry of Education, Science and Technology (in Korean).

Yoo, K.S. (1989), "Does Human Capital Investment Increase Inequality of Earnings in Korea?" *Journal of Economic Development*, 14(1), pp. 163-174.

Appendix

Figure 1.1: Histogram: Log Monthly Net Wage

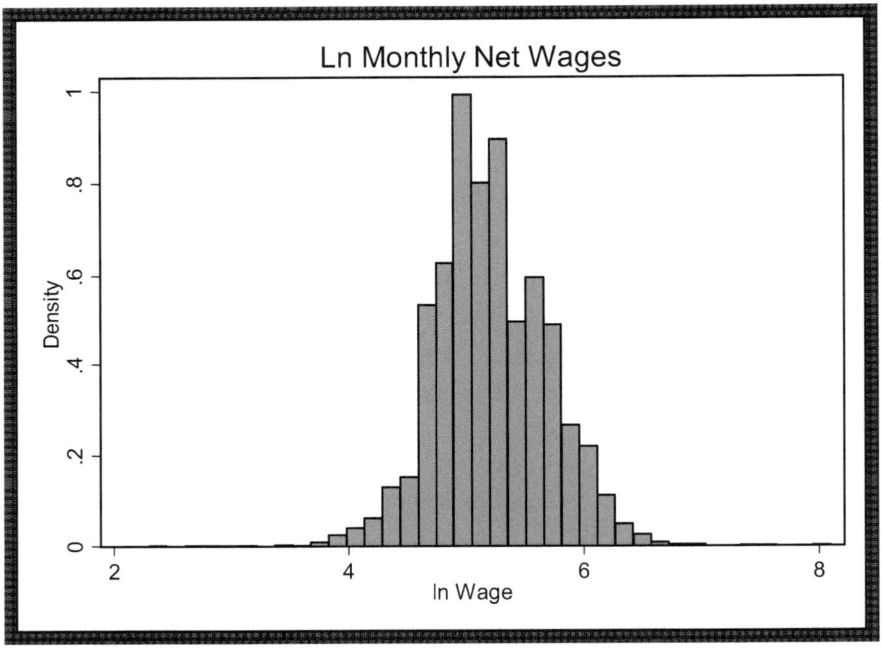

Table 1.3: Descriptive Statistics (6,649 observations males aged between 26 and 55)

Variable	Mean	Std. Dev	Min	Max
Ln wage	5.214498	.4749601	2.3	8.1
Real wage (Base year: 2000, Won)	1,848,955	1,018,252	90,300	28,005,800
Age	38.12667	7.745222	26	55
Experience dummies				Percent
less than 5years				10.14
5-9 years				12.08
10-13 years				12.25
14-18 years				10.82
19-23 years				13.69
24-28 years				11.03
29-33 years				11.64
34-40 years				9.09
41-48 years				8.14
49years and more				1.13
Tenure dummies				Percent
less than 1 year				10.89
1year				15.59
2 years				11.85
3 years				9.33
4 years				7.49
5-6 years				10.34
7-9 years				10.19
10- 13 years				9.66
14-19 years				9.04
20 years and more				5.63
School degree				Percent
Elementary school				3.42
Junior high school				8.79
Senior high school				40.35

Junior college				14.47
Bachelor				28.61
Master				3.70
Doctor				0.66
Completed job training last period				15.47
Job training in one's life time				30.46
Marital status				Percent
Single				21.89
Married				75.07
Separated				1.29
Divorced				1.47
Bereaved				0.27
Household head				78.61
Children age u.d. 18 dummy				61.10
Residence				Percent
Seoul				22.04
Busan				9.28
Daegu				6.54
Deajeon				2.89
Incheon				6.47
Gwangju				3.04
Ulsan				5.07
Gyunggi				21.41
Gangwon				1.22
Chungbuk				2.03
Chungnam				2.68
Jeonbuk				2.62
Jeonnam				2.12
Gyungbuk				3.93
Gyungnam				8.65

Jeju				0.02
Firm size				Percent
1-4 employees				10.03
5-9				10.94
10-29				17.39
30-99				17.17
100-299				12.04
300-499				4.47
500-999				4.87
1,000 and more				23.09
Firm type				Percent
Private company				87.96
Foreign company				1.53
Government invested, public business				5.48
Body corporate				4.44
The unattached				0.02
Civic-, religious, social groups				0.36
U.S. troops				0.05
Superintendent's office of apartment houses				0.02
Civil- and religious corporation				0.05
Social welfare organization				0.03
Dispatch worker service firm				0.02
The U.S. Army Headquarters				0.05
Educational institution				0.02
Regular worker				89.53
Part time				1.01
Temporary worker				Percent
Labor with non-temporary contract				92.28
Casual labor				3.81
Day labor				3.91

Overtime				63.04
Shift worker				13.13
Union				30.40
Year				Percent
2002				17.48
2003				17.41
2004				19.15
2005				22.57
2006				23.39

Note *The description of 202 industry dummies is omitted.*

Table 1.4: Descriptive Statistics for Wage Function Heckman 2 Step (779 observations for males aged between 26 and 55)

Variable	Mean	Std. Dev	Min	Max
Ln wage	5.417895	.4798191	3.91	7.38
Real wage (Base year: 2000, Won)	2,536,239	1,370,420	500,000	16,000,000
Experience dummies				Percent
less than 5years				13.22
5-9 years				21.18
10-13 years				19.38
14-18 years				17.72
19-23 years				11.30
24-28 years				10.01
29-33 years				3.98
34-40 years				3.21
Tenure dummies				Percent
less than 1 year				8.47
1year				13.22
2 years				9.50
3 years				8.73
4 years				7.06
5-6 years				10.40
7-9 years				12.97
10- 13 years				13.86
14-19 years				11.81
20 years and more				3.98
School degree				Percent
Elementary school				33.76
Junior high school				0.77
Senior high school				3.85
Junior college				12.07
Bachelor				42.62
Master				6.16

Doctor				0.77
Completed job training last period				23.36
Job training in one's life time				41.34
Marital status				Percent
Single				8.34
Married				90.63
Separated				0.39
Divorced				0.64
Household head				88.19
Children age u.d. 18 dummy				77.92
Residence				Percent
Seoul				26.70
Busan				2.95
Daegu				3.72
Deajeon				5.78
Incheon				1.80
Gwangju				4.11
Ulsan				4.62
Gyunggi				21.05
Gangwon				0.64
Chungbuk				2.05
Chungnam				4.49
Jeonbuk				4.62
Jeonnam				3.21
Gyungbuk				4.36
Gyungnam				9.88
Firm size				Percent
1-4 persons				6.55
5-9				7.06
10-29				13.99

30-99				16.69
100-299				12.45
300-499				5.13
500-999				5.78
1,000 and more				32.35
Firm type				Percent
Private company				84.98
Foreign company				2.31
Government invested, public business				7.83
Body corporate				4.62
Civic-, religious, social groups				0.13
U.S. troops				0.13
Regular worker				92.17
Part time				0.90
Temporary worker				Percent
Labor with non-temporary contract				96.02
Casual labor				2.44
Day labor				1.54
Overtime				53.40
Shift worker				14.12
Union				39.41
Year				Percent
2002				15.15
2003				14.63
2004				20.41
2005				23.36
2006				26.44

Note *The description of 202 industry dummies is omitted.*

Table 1.5: Effects of School Education and Job Training on Wages

Dependent Variable: Wage								
Independent	**OLS**		**Heckman 2 step**		**Random effects**		**Fixed effects**	
Elementary school	-.161670 -6.04	-.160268 -5.99	-.361944 -2.82	-.348542 -2.69	-.152610 -4.10	-.151480 -4.07	.092655 1.48	.090426 1.47
Junior high school	-.110880 -6.17	-.109976 -6.12	.042639 0.60	.045215 0.63	-.128794 -4.91	-.128662 -4.91	-.059160 -0.83	-.061793 -0.90
Junior college	.083744 6.15	.082268 6.06	.173739 3.78	.167905 3.62	.094129 4.60	.092502 4.54	.306828 1.56	.300148 1.53
Bachelor	.243269 15.62	.239338 15.39	.301736 7.08	.302453 7.03	.274646 12.14	.272326 12.06	.344841 1.46	.341879 1.45
Master	.550434 15.12	.544152 15.00	.710350 7.92	.707253 7.80	.543101 10.10	.538232 10.04	.675647 2.32	.661552 2.28
Doctor	.706501 8.56	.698900 8.52	.887564 3.70	.890110 3.63	.800411 6.02	.791186 5.97	3.591455 6.58	3.548966 6.51
Job Training in life	.043463 4.61	-	.030118 1.08	-	.030264 3.11	-	.018386 1.38	-
Job training Last period	-	.067875 5.69	-	.000531 0.02	-	.035723 3.67	-	.022809 2.01
Mill's lambda	-	-	.215809 3.16	.226734 3.33	-	-	-	-
Observations	6649	6649	1328 Cens. 549	1328 Cens. 549	6649 Groups 2545	6649 Groups 2545	6649 Groups 2545	6649 Groups 2545
R-sq/ Wald chi2	0.5704	0.5710	2317.08	2308.73	17379.27	26424.69	0.3882	0.3887

Note *t-statistics and z-statistics are presented in the 2nd line of cells in the table. Base variable (BV) of education: Completion of senior high school education. Base variable of job training: Absence of job-training.*

Table 1.6: Probit Estimation Results of Selection Function by Heckman 2 Step

Dependent Variable: Dummy Wage					
Independent	**Marginal Effect**	**z-statistics**	**Independent**	**Marginal Effect**	**z-statistics**
Year (Base variable (BV): 2002)			**Family Residence** (BV: Seoul)		
2003	.079086	1.79	Busan	-.248467	-4.32
2004	.036374	0.87	Daegu	-.020196	-0.29
2005	.025173	0.61	Deajeon	-.134483	-2.61
2006	.079563	1.94	Incheon	-.160578	-1.92
Experience (BV: Under 5 year)			Gwangju	.041804	0.57
5-9 years	.140381	2.39	Ulsan	.202615	2.06
10-13 years	-.058449	-0.99	Gyunggi	.035037	0.89
14-18 years	-.172602	-2.98	Gangwon	-.172657	-1.39
19-23 years	-.181990	-2.92	Chungbuk	-.076729	-0.84
24-28 years	-.187127	-2.88	Chungnam	-.128880	-2.10
29-33 years	-.258769	-3.67	Jeonbuk	-.058907	-1.03
34-40 years	-.175065	-2.00	Jeonnam	-.199271	-2.88
41-48 years	-.378320	-2.08	Gyungbuk	-.169008	-2.85
Education (BV: Senior High School)			Gyungnam	.025246	0.47
Element. school	-.011023	-0.12	**Family Characteristics** (BV for marital status: Single)		
Jun. high school	-.035941	-0.61	Married	.027079	0.33
junior college	-.009163	-0.21	Separated	.106172	0.68
Bachelor	.054741	1.51	Divorced	.025299	0.13
Master	.065771	0.95	Household head	.155456	2.26
Doctor	-.091531	-0.47	Children	.018364	0.48
Job Training (BV: No Job training)			**Household (hh) Finance**		
Training in life	.146763	5.03	Ln finan. wealth	.003645	0.39
House Ownership (BV: Own House)			Ln debt	-.0957296	-8.95
Lease of a house on a deposit basis	-.020401	-0.65	Ln non labor income	-.011156	-1.41
Lease of a house with monthly rent	-.033729	-0.54	Ln sustenance allowance for parents	.029341	2.37
Uncertain dwelling	-.186926	-2.44			
Log likelihood	-851.37025				

Table 1.7: Effects of Experience on Wages

Dependent Variable: Wage		
Experience (year)	**Heckman 2 step (z-statistics)**	**Random effects (z-statistics)**
5-9	.075129 1.55	.103631 6.43
10-13	.259247 4.55	.151155 7.63
14-18	.331081 4.88	.215390 9.34
19-23	.346283 4.39	.218048 8.23
24-28	.289502 3.61	.240857 7.85
29-33	.278854 2.57	.198937 5.32
34-40	-.049691 -0.47	.125089 3.05
41-48	-	.060369 0.72
49-	-	.249614 2.08
Experience	.047017 6.96	.017579 7.32
Experience squared	-.001172 -6.72	-.000297 -5.27

Note *Base variable: less than 5 years. z-statistics are presented in the 2nd line of cells in the table.*

Table 1.8: Effects of Tenure on Wages

Dependent Variable: Wage		
Tenure (year)	**Heckman 2 step (z-statistics)**	**Random effects (z-statistics)**
1	.040683 0.89	.035393 2.67
2	-.062218 -1.34	.053850 3.81
3	-.058041 -1.20	.066833 4.35
4	-0.056457 -1.09	0.080611 4.65
5-6	.018961 0.37	.101127 5.49
7-9	.002727 0.05	.144963 6.99
10-13	.176292 3.08	.215909 9.13
14-21	.134722 2.25	.314639 11.62
22 and more	.536355 5.21	.380069 9.74
Tenure	-.000561 -0.10	.018243 6.91
Tenure squared	.000706 3.09	-.000150 -1.33

Note *Base variable: less than 1 year. z-statistics are presented in the 2nd line of cells in the table.*

Figure 1.2: Effects of Experience on Wages

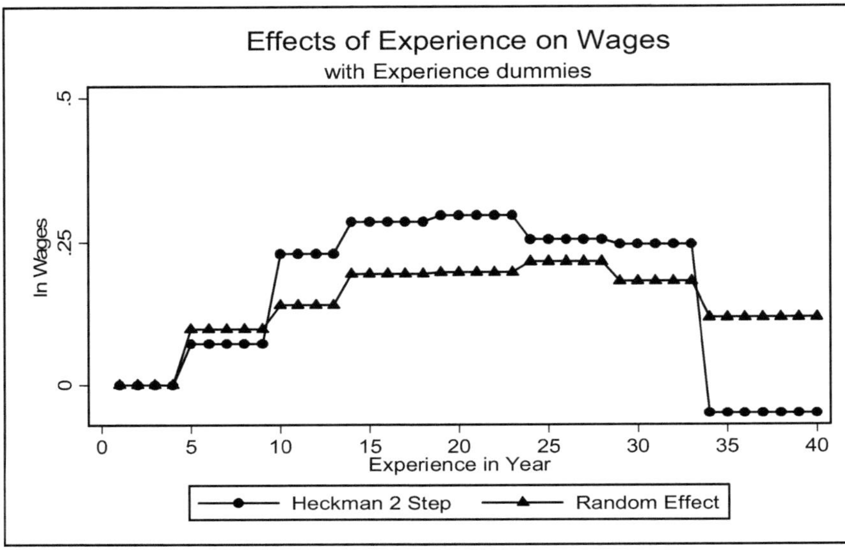

Note *Base Variable: less than 5 years.*

Figure 1.3: Experience-Wage Profiles

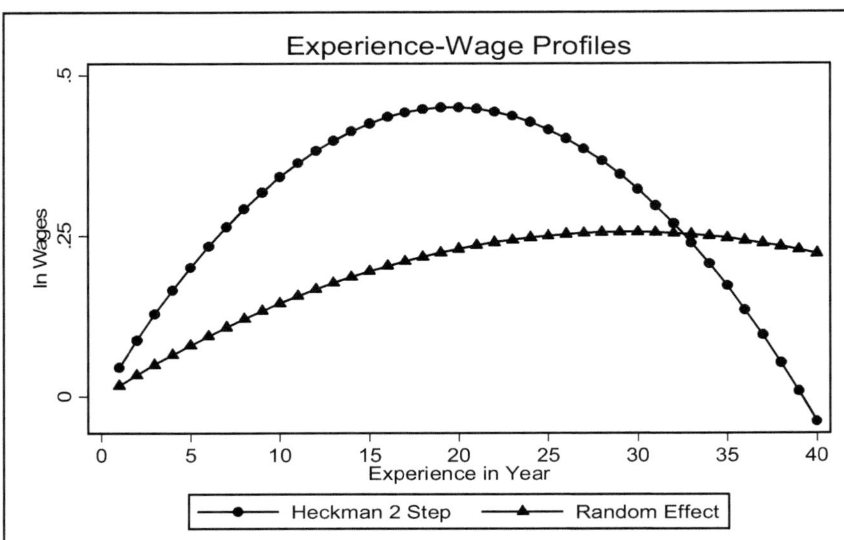

Figure 1.4: *Effects of Tenure on Wages*

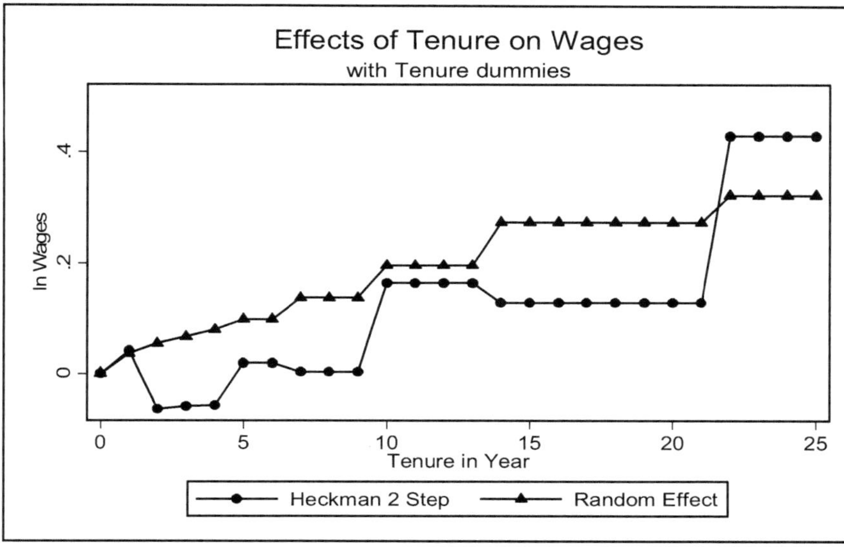

Note *Base Variable: less than 1 year.*

Figure 1.5: *Tenure-Wage Profiles*

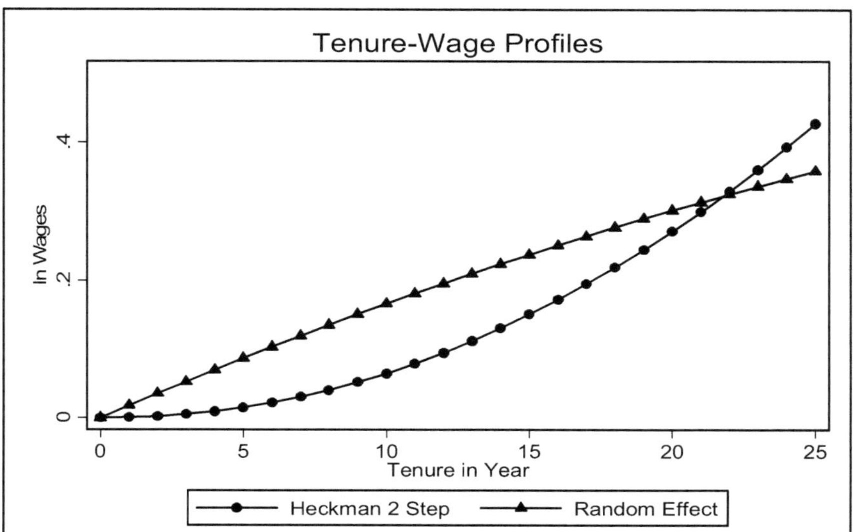

Table 1.9: Effects of Family Background on Wages

Dependent Variable: Wage		
Family background	**Heckman 2 step (z-statistics)**	**Random effects (z-statistics)**
Married	.198627 2.50	.099971 4.74
Separated	.458127 1.82	.059323 1.17
Divorced	.276827 1.18	.068510 1.83
Bereaved	-	.063284 0.64
Household head	.061829 1.02	.095322 4.70
Having children	.038597 0.96	.029971 2.38

Note *Base variable of marital status: Single. z-statistics are presented in the 2nd line of cells in the table.*

Table 1.10: Effects of Working Contracts, Working Conditions and Trade Unions on Wages

Dependent Variable: Wage		
Working contract, working condition and trade unions	**Heckman 2 step (z-statistics)**	**Random effects (z-statistics)**
Regular labor (BV: Irregular labor)	.173947 3.22	.090512 5.37
Casual labor (BV: Non-temporary labor)	-.080597 -1.04	-0.082413 -2.86
Day labor (BV: Non-temporary labor)	.196961 1.98	-.084675 -3.10
Part time	.097848 0.81	-.089681 -1.30
No overtime	-.019116 -0.87	-.031919 -4.61
Working Shift	.008487 0.22	-.037006 -2.32
Having Union	.108885 3.39	.039590 3.56

Note *BV: Base variable. z-statistics are presented in the 2nd line of cells in the table.*

Table 1.11: Effects of Firm Size (Number of Employees) on Wages

Dependent Variable: Wage		
Firm size (Number of employees)	**Heckman 2 step (z-statistics)**	**Random effects (z-statistics)**
5-9	.140638 2.27	.053828 3.13
10-29	.290093 4.95	.118147 6.65
30-99	.220897 3.72	.129793 7.03
100-299	.242697 3.72	.139622 6.96
300-499	.329575 4.15	.175527 7.18
500-999	.490237 5.76	.198710 8.34
1,000 and more	.490873 6.66	.235220 10.13

Note *Base variable: Firm with 1-4 employees. z-statistics are presented in the 2nd line of cells in the table.*

Table 1.12: Effects of Family Residences on Wages

Dependent Variable: Wage		
Residence of Family	**Heckman 2 step (z-statistics)**	**Random effects (z-statistics)**
Busan	.053862 0.66	-.093217 -3.90
Daegu	-.292220 -5.24	-.170501 -6.91
Deajeon	-.146581 -2.62	-.100879 -2.99
Incheon	-.112025 -1.35	-.131865 -5.32
Gwangju	-.167692 -2.67	-.108611 -3.12
Ulsan	.045785 0.66	.004058 0.14
Gyunggi	-.002183 -0.06	-.051475 -3.04
Gangwon	-.131041 -0.75	-.252145 -4.32
Chungbuk	-.226565 -2.87	-.076096 -1.98
Chungnam	-.1129404 -1.84	-.0566348 -1.76
Jeonbuk	-.156189 -2.71	-.121761 -3.56
Jeonnam	.030389 0.42	-.060222 -1.05
Gyungbuk	-.156511 -2.55	-.147697 -4.49
Gyungnam	-.161384 3.63	-.095636 -4.23
Jeju	-	.480272 2.47

Note *Base variable: Seoul (Capital city of Korea). z-statistics are presented in the 2nd line of cells in the table.*

Table 1.13: Effects of Firm Types on Wages

Dependent Variable: Wage		
Firm type	**Heckman 2 step (z-statistics)**	**Random effects (z-statistics)**
Foreign company	.057380 0.77	.017659 0.59
Government invested, public business	-.096596 -1.98	-.005247 -0.23
Body corporate	.041891 0.67	.036203 1.72
The unattached	-	.112468 3.01
Civic-, religious, social groups	-.438047 -1.38	-.184772 -2.16
U.S. troops	.670977 1.81	-.010040 -0.05
Superintendent's office of apartment houses	-	.138963 4.62
Civil- and religious Corporation	-	-.344522 -2.36
Social welfare organization	-	.237018 1.44
Dispatch worker service firm	-	-.053367 -0.73
The U.S. Army Headquarters	-	.111144 0.71
Educational institution	-	.384392 3.21

Note *Base variable: private company. z-statistics are presented in the 2nd line of cells in the table.*

2. Effects of School Education and Job Training on Earnings in Korea and Germany. A Comparative Study

Abstract

Researchers in Korea have advanced the hypothesis that the earnings gap between university graduates and those with lower levels of education is higher in Korea than in Germany due to German job training reducing the earnings gap between educational levels. This lessens competition for university entrance in Germany. This study estimates earnings differentials between educational levels and job training for both countries using the 2002-2007 dataset comprised of male employees taken from the 'Korea Labor and Income Panel Study' and 'German Socio-Economic Panel'. Utilizing quantile regression method, this study repudiates the hypothesis: Earning differentials between education levels are higher in Germany than in Korea, although German job training compensates for these differentials more strongly than Korean job training.

Keywords: School education, Job training, Human capital, Earnings, Korea, Germany

2.1. Introduction

According to educational indicators supplied by the OECD, the rate of entry to university for the relevant aged-cohort in the population in 2006 was 35% in Germany. In the same year, this entry rate reached 59% in Korea (OECO (2008)). The relatively high entry rate to university in Korea has its downsides however. While the unemployment rate of university graduates was about 1% in Germany, the percentage of the unemployed who graduated from a university among all the registered unemployed was 12.6% in 2000, 17.1% in 2004, 19.5% in 2007 and 22.04% in 2009 in Korea (Kolja Briedis (2009), Korean Statistical Information Service (2010)). Higher educated people suffer more seriously from unemployment. Nevertheless, the educational fever to enter university still keeps growing in Korea. Since 2004, more than 80% of all high school graduates have gone on to university or junior college. In 2008, this reached 83.8%.

When considering the inefficiently high competition for university entrance, Korean education experts are concerned about the deficit of vocational education at the senior high school level in Korea and have shown interest in the German dual system. Korean politicians have tried to introduce a vocational training system at school along German lines: 21 'meister senicr high schools', whose names are taken from the German '*Meisterschule*', were proposed and opened in March 2010 (Ministry of Education, Science and Technology (2009)). With this project, the Korean Government hopes to reduce inefficient competition for university entrance and supply a trained labor force to the manufacturing industry. But, on the other hand, Koreans are deeply prejudiced against people without a university degree. They consider a university degree to be a necessary condition for a successful working/social life. I question whether an introduction of German educational formalities into the Korean school system could help to guarantee the successful development of (vocational) education in Korea. Given these concerns, a comparative study of the education and job training effect on the labor market in Korea and Germany becomes more important.

In Korea, the German (vocational) education system has been frequently considered. Some educational and labor market institutes have been interested in the German education and vocational training system, related policies and reform plans as a possible reference for education and labor market policy-making in Korea. However, in these studies[13], there were no analyses concerning the eco-

---

13 There are several international comparisons of education and vocational training systems published by the Korean Educational Development Institute, the Korea Labor Institute and the Korea Research Institute for Vocational Education & Training, in which the German system is presented. The Human Resources Development Service of Korea translated some German texts, which are relevant for any study of vocational training, for example, *Das Berufsbildungssystem in der Bundesrepublik Deutschland* (1994) published by the European Centre for the Development of Vocational Training (CEDEFOP), *Reformprojekt Berufliche Bildung-Flexible Strukturen und moderne Berufe* (1998) published by the Federal Ministry of Education and Research (Bundesministerium für Bildung, Wissenschaft, Forschung und Technologie); *The role of national qualificaions systems in promoting lifelong learning - Background report for Germany* (2003) by Jochen Reuling and Georg Hanf. Furthermore, some German researchers have written about German vocational education and training in the Korean journals, for instance, Lothar Funk (2005) "Situation and Vision of the Vocational Educational Training System in Germany," *International Labor Brief*, 3(6). Korean researchers have also published articles concerning the German vocational education system for example, Ki Hong Kim (1998) "Die neue Entwicklungstendenz der Berufsbildung in Deutschland und Hinweise auf unsere Berufsbildungspolitik", *Koreanisch-Deutsche Forschung der Erziehungswissenschaft*, 3(1).

nomic returns due to German education and vocational training. In 1998, the Korea Labor Institute launched the 'Korean Labor and Income Panel Study (KLIPS)', which is comparable to the 'German Socio-Economic Panel (GSOEP). This strongly supports a comparative analysis of the economic returns attributable to (vocational) education in Korea and Germany, which is the aim of this study. This attempt contributes to an expansion of the method of comparative study between Korea and Germany into the sphere of empirical economics.

The main focus of this study is: why is a university education in Korea considered to be so much more important than is the case in Germany? Koreans often advance the hypothesis that the earnings differential between university graduates and workers with a lower level of education is larger in Korea than in Germany, where vocational training at school and on-the-job training for adults are well developed. For example, Kim (2006), who is one of the Korean specialists in the German job training system, asserts that the German vocational education - dual system - enables workers with a low formal education to qualify as technical experts and thus reduces the earnings gap between university graduates and senior high school graduates, thereby lessening inefficient competition for university entrance. This study adopts this hypothesis as a possible answer to the question and seeks to demonstrate the effects of education and job training on earnings in Korea and Germany. I intend to confirm whether well-developed German vocational education and training reduces the earnings gap between those with lower education and higher level education when compared to the situation in Korea. It focuses on earnings differences between senior high school graduates and university graduates.

Following this introduction, I move on to a consideration of the differences in school educational systems in Korea and Germany in the second section of this paper. Then, the model and method will be described in the third section. I will explain the data and variables and present descriptive statistics in the following section, leading to the estimation result and a final conclusion.

2.2. Difference of School Education and Job Training between Korea and Germany

2.2.1. School Systems in Korea and Germany

The educational system in Korea is composed of 6 years elementary school, 3 years junior high school, 3 years senior high school, followed by either 2 years junior college or 4 years university. After having completed 6 years of elementary school (ISCED-97 level 1), children move on to a junior high school (ISCED-97 level 2). The 9 year period of education from elementary school to junior high school is compulsory. After 3 years of junior high school education, pupils take the entrance examination to the senior high school (ISCED-97 level 3). If they pass this examination, they can enter either academic senior high school or vocational senior high school. An academic senior high school requires a better mark from the entrance examination than a vocational senior high school. After graduation from senior high school, pupils go on to university or junior college (ISCED-97 level 5[14]). For entrance to a university or a junior college, pupils have to attain an appropriate mark in the entrance exam. Entrance into the tertiary educational institutions is permitted to all senior high school graduates. Not only pupils who have completed an academic but also graduates from vocational senior high school can enter university if they attain an appropriate mark in the university entrance exam. About 67% of graduates from vocational senior high school go on to junior college (ISCED-97 level 5B) nowadays. The remaining pupils of vocational senior high school undertake about 1-6 months of practical vocational training in a company and look for a job. After 4 years of bachelor study at university (ISCED-97 level 5A-1st degree), Koreans usually complete their schooling. About 13% of university graduates take a master's course (ISCED-97 level 5A-2nd degree), and about 1.4% of them proceed to study in a doctoral course (ISCED-97 level 6). A master's course takes 2 years and a doctoral course more than 3 years (Center for Educational Statistics (2009)).

14 ISCED- 97 level 5 comprises the first stage of tertiary education. Programs at level 5 are subdivided into 5A, programs that are largely theoretically-based and are intended to provide sufficient qualifications for gaining entry into advanced research programs and professions with high skills requirements, and into 5B, programs that are generally more practical/technical/occupationally specific than ISCED 5A programs. ISCED-level 5A can be sub-classified by the theoretical cumulative duration of programs into bachelor course (5A-1st degree) and master course (5A-2nd degree). See OECD (1999).

Table 2.1: School Systems in Korea and Germany

Class	Korea/*ISCED-97 level*		Germany/*ISCED-97 level*			
19	(3 years)** Doctor's course/*6**				Promotiom/*6*	
17	(2 years) Master/*5A*				(2 years) Master/*5A*	
14					(3 years) Bachelor/*5A*	(5-6.5 years) Diplom, Magister/*5A* <u>Universität</u>
13	(4 years) Bachelor/*5A* <u>University</u>	(2 years) Junior college/*5B*	(1-3 years) Fachschule/*5B*	(3-4 years) Fachhochschule/*5A*		
11				(2-3 years) Fachoberschule Berufsoberschule Berufsfachschule/*3*	(2-3 years) Gymnasiumoberstufe Fachgymnasium/*3*	
10	Academic/vocational Senior high school/*3*		Berufschule/*3*			
7	Junior high school/*2*					
5			Hauptschule/*2*	Realschule/*2*	Gymnasium/*2*	
1	Elementary school/*1*			Grundschule/*1*		
0	Kindergarten/*0*			Kindergarten/*0*		

Sources: *1) Kim, A.C. et al. (2006), 2) OECD (1999), 3) Hámori, S. (2007)*

Note ** The italic numbers in the table show us the ISCED-97 level of school education. ** The years in parentheses indicate duration of school education. Other schooling durations can be recognized by the numbers of classes.*

After completing four years *Grundschule* (ISCED-97 level 1) in Germany, children choose their schooling route: *Hauptschule*, *Realschule* or *Gymnasium* (ISCED-97 level 2) at the lower secondary level of education. According to school law in the German States, this *Grundschule* education can be prolonged up to the 6th class or, alternatively, the first 2 years of schooling at the lower secondary educational level offers an orientation course which will definitively determine the pupils further schooling route. 5 years *Hauptschule* offers a general education with a vocational orientation. A successful completion of the *Hauptschule* enables vocational education at the upper secondary level, but does not further an academic education. Graduating after 6 years of *Realschule* allows pupils to go on to schools that provide occupational qualifications or qualifications for a higher level of general education. The *Gymnasium* is the academic high school at the lower secondary level of education, in which pupils prepare for the higher level of general secondary school, *Gymnasiale Oberstufe* (ISCED-97 level 3). The comprehensive school, *Gesamtschule*, combines all these three tracks in the secondary level of education. Unlike Korea, where pupils decide their schooling route at the age of 15, German pupils choose their schooling route at the age of 10. The German education focuses on ones' specification from an early age.

The upper secondary level of education demonstrates a great difference in schooling systems between Germany and Korea (see Tables 2.1 and 2.2). The dual system for *Hauptschule* graduates lasts 3 years. That offers an apprenticeship at a company during the general education at part-time vocational school, namely *Berufsschule* (ISCED-97 level 3). Attaining vocational experience in the meantime, graduates from the *Berufsschule* can earn further qualifications through an advanced vocational education lasting for 3 years at the *Fachschule* (ISCED-97 level 5B). The dual system for *Realschule* graduates is undertaken in *Berufsfachschule* (ISCED-97 level 3). *Berufsfachschule* also grant admission to *Gymnasium* graduates. This course offers a two or three year full-time vocational education and provides pupils with vocational training as well as general education. The two-year upper secondary schools, *Fachoberschule* and *Berufsoberschule* (ISCED-97 level 3), offer the *Fachhochschulreife* to their participants. The upper secondary level of education in Germany is much more diversified and oriented towards vocational practices than in Korea.

Table 2.2: Upper Secondary and Post-Secondary Non-Tertiary Education in Korea and Germany

Korea				Germany			
School	ISCED -level	Duration in year	Cum. school years	School	ISCED -level	Duration in year	Cum. school years
Academic senior high school	3A	3	12	Fach-oberschulen	3A	2	12
				Berufs-fachschulen	3A	2-3	12
				Fachgynasium /Gynasium-oberstufe	3A	3	13
Vocational senior high school	3C	3	12	Berufsfach-schulen	3B	1-2	11
				Duales System/ Berufsschule	3B	3	13
				Duales System/ Berufsschule	4A/4B	3	16
				Fach-oberschulen	4A	1	14
				Berufs-/ Technische Oberschulen	4A	2	15

Source: *OECD (1999).*
Note *ISCED 3A: Programs at Level 3 designed to provide direct access to university. ISCED 3B: Programs at Level 3 designed to provide direct access to junior college or to German trade-/technical school. ISCED 3C: Programs at Level 3 designed to prepare students for direct entry into the labor market, although they also provide access to ISCED 4 programs or other ISCED 3 programs.*

Fachhochschulen and *Universität* (ISCED-97 level 5A) comprise the first stage of tertiary education. *Fachhochschulen*, where education lasts 3 or 4 years, focuses more on vocational practice than university. Educational courses at German universities, which usually take 5 to 6.5 years, are divided into *Diplom* and *Magister* courses. According to the Bologna accords in 1999, European countries succeeded in unifying their academic degree standards and quality assurance standards according to a common level. The main agreement was the introduction of standard bachelor's and master's courses at European universities. In 2005, 27% of all study courses at German universities constituted bachelor and master according to the Bologna accords. A bachelor's course lasts 3 years and a master's course 2 years in Germany. Under certain conditions, students can advance to a doctoral course at the second stage of tertiary education (ISCED-97 level 6). Unlike Korea, university entrance is not excessively demanding for students, but completing university study is not always possible for students in Germany. Students have to demonstrate their competence at specific intervals during their study: This leads to a high 'drop-out' rate from university. In Korea, entrance into a university is highly demanding, but after entrance, generally, every student is capable of completing his/her university education.

2.2.2. Vocational Training at the Upper Secondary Education in Korea and Germany

German Vocational Schooling The German dual system is conducted by means of co-operation between the Federal Government, individual States and other social partners. These social partners consist of: chambers of manufacture, commerce, craftsmen, agriculture, freelancers and other industries, along with employers and trade unions. The laws - *Berufsbildungsgesetz* (1969) and *Berufsbildungsreformgesetz* (2005) - regulate vocational training. The Federal Institute for Vocational Training (*Bundesinstitut für Berufsbildung*) provides rules and programs for vocational training. The Standing Conference of the Ministers of Education and Cultural Affairs of the States (*Ständige Konferenz der Kultusminister*) decrees curricula for relevant occupational instruction (*Rahmenpläne*) at vocational schools. Each State develops curricula for general instruction (*Lehrpläne*). The German Federal Employment Agency (*Bundesagentur für Arbeit*) is responsible for counseling, placement and aid of vocational training for pupils and establishments. Social partners participate in developing a norm for vocational training, providing references, coordination between schools and establishments, advice and inspection of vocational training in establishments. Other functions of social partners include: control examinations and the assignment of degrees as well as certification, and negotiations for the supply of vocational training positions, working condition and wages of trainees (Hippach-Schneider et al. (2007)).

Entrance into the German dual system is permitted to graduates from all kind of lower secondary level schools, such as: *Sonderschulen, Hauptschulen, Realschulen, Gesamtschulen, berufliche Schulen* and *Gymnasien.* 3 years of vocational education in this dual system is carried out in both establishments and vocational schools. The purpose of vocational education in these establishments is to help pupils to achieve the qualifications necessary for skilled occupational tasks. Successful completion of this training enables the pupil to become a specialist in any of 346 accredited occupations (The Federal Institute for Vocational Training (2006)). Pupils participate for 3 to 4 days a week in vocational training in an establishment and are taught for 1 to 2 days in vocational school. Firms, which train pupils, are responsible for imparting competence in each occupation requiring formal training according to the national standard of occupational skill states. Upon completion of vocational training in an establishment, pupils have to pass an examination demonstrating their occupational competence. A vocational school in the dual system imparts basic or specialized vocational education in addition to the teaching of further general education. A minimum of 12 teaching hours is required: 8 hours for occupational instruction and 4 hours for general subjects such as German, social studies, economics, religion and foreign languages. To gain certificates from this school, pupils have to acquire a certain mark. In Germany, about 52.5% of the relevant age cohort in 2004 acquired an accredited job, requiring training in the dual system. Pupils can acquire advanced vocational training on the job and, moreover, attain a higher level of education under certain conditions. In 2005, about 20% of matriculating students at university acquired vocational training in the dual system.

Korean Vocational Schooling Vocational education at the upper secondary level is regulated by the Fundamentals of Education Act (1998) and the Law concerning Primary and Secondary School Education (2005) in Korea. Their curricula are centrally decided by the Minister of Education, Science and Technology. In the application of these curricula at vocational schools, the Superintendent of Educational Affairs in Cities and States provides an administrative guide. According to the curricula, all pupils learn 10 general subjects up to their 10th schooling year: Korean, ethics, society, math, science, technology, home economics, gymnastics, music, art and English. In the 11th and 12th schooling year, pupils begin to specialize according to their educational route. At this point, pupils at vocational senior high school take at least 82 units of lessons for their vocational specialties between the 11th and 12th class. Specialties at the vocational school consist of 58 majors (Ministry of Education, Science and Technology (1997)). Pupils at a vocational school should study vocational training in an establishment run in accordance with the Law promoting Vocational Education and Training (1997). Vocational training at such schools is additionally regulated by the Rule concerning Vocational Training of Schools (1999). This training is planned and performed by a committee in which the central government, the regional governments, representatives of job training institutes, employers and employees all participate.

According to these rules, directors at vocational schools choose appropriate establishments, where their pupils undertake vocational practice. The regional committee of job training, constituted by the regional government, representatives of job training institutes, employers and employees in their region, can advise school directors in their choice of establishment. Employers select trainees who they consider to be suitable for their establishment. There are no master curricula or national standard controlling the quality of vocational practice. The practice lasts between 34 hours and 6 months. This duration varies case by case for each pupil. If necessary, a school director can prolong it by one year at their own discretion. Vocational practice can replace lessons at school, for example, with between 5 and 6 months practice held to be equivalent to 6 months of lessons at school. Trainees do not receive any kind of certificate after their vocational practice. Through this practice, pupils can only become acquainted with the atmosphere of an establishment, but can not acquire a skill. Among all senior high school graduates[15], those who completed vocational senior high school comprised 31% and those who directly participated in the labor market after graduating school comprised about 11% in 2004 (Korean Educational Development Institute (2005), pp. 31). The numbers of pupils who take vocational education are much lower in Korea compared with in Germany. While the German dual system plays a central role in labor supply, vocational education in Korea does not play a significant role in the labor market.

2.2.3. Job Training after Schooling in Korea and Advanced Job Training in Germany

Advanced Job Training in Germany About 12% of the population between the ages of 25 and 64 take part in informal job-related continuing education and training and the expected hours of that were 398 hours per annum in Germany (OECD (2007), pp. 353). The Federal Ministry of Education and Research (*Bundesministerium für Bildung und Forschung*) reported that about 26% of the population aged between 19 and 65 participated in advanced job training in 2003 (Kuwam (2006)). Nation wide standardized promotional training, refresher training and retraining, and on-the-job training by a firm as well as job training on one's own initiative make up what is termed advanced job training. In 2002, about 68% of participants in advanced job training undertook on-the-job training and about 25% off-the-job training. Around 13% participated in both of them, on-the-job and off-the job training (Beicht (2006)). Advanced job training at school is carried out on a part time or full time basis at the *Fachschulen* or special *Meisterschulen* (ISCED level 5B), where *Meister* or technicians obtain their qualifications. These schools require participants to possess an accredited occu-

---

15 The ratio of upper secondary school graduates (senior high school graduates) to the population at the typical age of graduation was about 93% in 2005 (OECD (2007), pp. 42).

pational qualification and more than one year of occupational experience in the relevant area, or graduation from a *Berufsfachschule* and more than 5 years corresponding job experience. Completion of such advanced job training at school is achieved when an examination is passed according to the federal state law. There are also many short term courses in the area of advanced job training, especially in the area of refresher training, for example a 3-day course.

Job Training after Schooling in Korea The first Law of Job Training was passed in 1967, at the dawn of industrialization in Korea. With the advent of industrialization plans (i.e. the five-year programs for economic development in Korea[16]), the demand for skilled labor expanded rapidly. The Vocational Training Center (1968) and the Korean-German cooperated Vocational Training Center (1971) were established at this time. The government began to promote on-the-job training through a subsidy. Between 1967 and 1973, the number of workers who completed a vocational training program reached 17,200 (Park (2007)). In 1974, the Special Law for Vocational Training was enacted. According to this law, firms with more than 200 employees were obligated to conduct on-the-job training. As a result, the number of firms carrying out on-the-job training increased from 30 in 1974 to 476 in 1976. Participants in on-the-job training also increased from 12,900 in 1974 to 96,800 in 1976 (Suh (2002)). Such a job training policy effectively supported the Heavy and Chemical industrialization of Korea in the 1980's.

In the 1990's, the level of school education was rapidly enhanced in Korea. Entrance rate to junior college and university was 23.7% for all senior high school graduates in 1980. This then rose to 51.4% in 1990, 68% in 2000 and to more than 80% recently. At the same time, Korean job training, which was strongly promoted by the government, was weakened. Job training was mainly necessary for unemployed or older people and for people in lower income groups. In 1997, the pre-existing law for vocational training was repealed, and the Promotion Law for Job Training of Labor, which diminished the role of the government and promoted firm's and employee's initiatives, was enacted. Although this law supports job training in several ways, job training has not been revitalized yet. In 2002 and 2003, the participation rate in informal job training for all adults was 4.67% and 5.13%. Among these, 50% of participants had received job training of 5 days or less (Lee et al. (2006), pp. 37). Increasing levels of education in the 1990's seem to have had a negative effect on job training activities in Korea. In

16 The five-year programs for economic development were planned and accomplished in 7 rounds between 1962 and 1996.

response to this situation, educational and labor market experts in Korea are critically reviewing the current development of job training.

2.3. Model and Methods

This study is based on the Mincer earnings model (1974). Estimating the earnings effects of education and job training by means of OLS, the Mincer earnings function, however, reveals some problems: Measurement error, omitted ability bias and selection bias. If an appropriate dataset or variables are available, we can attempt to remove such biases using several techniques, for example: using a proxy variable or an instrumental variable, employing the fixed effects method, the Heckman 2 step method or the control function method etc.[17] However, in the case of a comparative study, which compares different survey data, finding the same variable, which would enable the correction of such biases, is not always possible. Ashenfelter and Zimmerman (1993, 1997), however, in reference to this problem demonstrated that the omitted ability bias and the measurement error bias from schooling canceled each other out by means of OLS. Considering their study, the main problem in estimating the earnings effect of school education using the Mincer earnings function would then be the selection bias.

This study uses the quantile regression and fixed effects methods. Concerning job training, the fixed effects method corrects the omitted ability bias by reference to this variable. Concerning the measurement error, people, in general, do not consciously overstate their job training experience in answering a questionnaire. So, measurement error would not be a problem attributable to the job training variable. A particular problem in estimating the job training effect on earnings using the fixed effects method is that it does not exclude the selection bias from any job training variables. Thus, the estimation result of this study suffers from selection bias, using the quantile regression and fixed effects methods. But the purpose of this study is primarily a comparison of differences in educational and job training effects on wages in Korea and Germany. Although these methods do not totally remove such biases, the estimation results can certainly demonstrate the relative differences in educational and job training effects on earnings between both countries.

---

17 See Griliches (1979); Heckman (1979); Ashenfelter and Krueger (1992); Ashenfelter and Zimmerman (1993); Behrman et al. (1980); Blackburn and Neumark (1991); Ashenfelter and Zimmerman (1997); Garen (1984), and Heckman and Robb (1985).

Quantile Regression An advantage of using quantile regression (Koenker and Bassett (1978), Koenker (2005)) is its robustness with respect to larger outliers in estimating at the median rather than ordinary least squares regression to estimate at the mean. Another advantage of quantile regression is the fact that any quantile can be estimated. The least squares regression curve gives a grand summary for the averages of the distributions corresponding to a vector of covariates. Quantile regression, however, can additionally present several different regression curves corresponding to the various percentile points of the distributions and thus offer a more complete picture of the distribution (Koenker and Hallock (2001)). So, unlike the least squares regression, these techniques allow the study of effects attributable to each of the covariates, for example, educational levels and job training, throughout the whole distribution as well as estimating the effect of workers' heterogeneity on earnings (Machado and Mata (2001)). Using quantile regression, this study analyzes not only the difference in earnings effects between education levels and job training (i.e. 'between-skill-levels earnings differentials'), but also presents a comprehensive picture of the effects of education and job training on earnings (i.e. 'within-skill-levels earnings differentials'). The quantile regression model is formulated as:

$$y_i = x_i'\beta_\theta + \mu_\theta, \quad Quant_\theta(y_i|x_i) = x_i'\beta_\theta \quad i=1,\ldots,n,$$

where y_i is the regression's dependent variable, x_i is a $K \times 1$ vector of covariates, μ_θ is an error term and β_θ is the vector of parameters to be estimated. The subscript *i* indexes the individuals in the sample, *i*=1,..., *n*. $Quant_\theta(y_i|x_i) = x_i'\beta_\theta$ denotes the θth conditional quantile of y_i given covariate x_i. The θth regression quantile can be computed by

$$\min_{\beta \in R^K} \left\{ \sum_{i:y_i \geq x_i'\beta} \theta |y_i - x_i'\beta_\theta| + \sum_{i:y_i < x_i'\beta} (1-\theta) |y_i - x_i'\beta_\theta| \right\},$$

minimizing the sum of absolute weighted residuals (Koenker and Bassett (1978), p. 38). $x_i'\beta_\theta$ describes the θth quantile of the earnings of individual *i* given covariates. This study primarily estimates earnings effects of educational levels and job training by means of OLS and the quantile regression at five quantiles of the log earnings distribution - at the 10th quantile, at the 25th quantile, at the median, at the 75th quantile and at the 90th quantile.

Fixed Effects Method The fixed effects method guarantees the acquisition of consistent estimators in the presence of time invariant unobserved omitted variables such as the omitted ability bias. In order to remove the unobserved effect, this method utilizes the difference between the function with the unobserved effects and the first averaging equation. Estimates obtained from this transformed function demonstrate, then, the difference in earnings from their average attributable to changes of education levels and job training. Concerning the education variable, we usually consider the effect of an adult education on earnings using the fixed effects method. According to the study by Ashenfelter and Zimmerman (1993, 1997), the fixed effects method could intensify the measurement error bias attributable to education due to the removal of omitted ability bias. Concerning job training, however, this method effectively presents the earnings effect of job training removing any time invariant omitted ability bias. Using this method, any estimation result will provide us with the earnings differentials due to a change of education level and job training for male adults between the ages of 18 and 55 in Korea and Germany.

2.4. Data and Descriptive Statistics

2.4.1. Data

To estimate the effects of school education and job training in Korea and Germany, this study uses the 2002 to 2007 data concerning male employees aged between 18 and 55 taken from two panel surveys: 'Korean Labor and Income Panel Study (KLIPS)' and 'German Socio-Economic Panel (GSOEP)'.

The 'Korean Labor and Income Panel Study (KLIPS)' is published by the Korea Labor Institute (KLI). KLIPS is a longitudinal survey of the labor market and income activities of households and individuals living in cities and urban areas. This survey is the first panel data source for the microeconomic analysis of labor market activities and transitions in Korea, thereby contributing to the development and evaluation of labor market policies. KLIPS is comparable to a set of successful longitudinal surveys in designing and managing, conducted in industrialized countries, such as NLS (National Longitudinal Surveys), NLSy (National Longitudinal Survey of Youth), and PSID (Panel Study of Income Dynamics) in the USA, SLID (Survey of Labour and Income Dynamics) in Canada, BHPS (British Household Panel Survey), and GSOEP (German Socio-Economic Panel), among many others. 5,000 households, constituting 13,321 household members in 7 big cities and urban areas in 8 provinces, participated in the survey in 1998 when the survey was launched. Respondents comprised 4,012 employees, 2,415 self-employed individuals and 7,315 non-employed people. The panel takes into account a variety of variables concerning social and economic backgrounds: Family, housing, income, consumption, education, job training, working contract, vocations, working conditions, job-seeking experience, mobility in the labor market, health, marital status, job contentment, etc. Income is divided into wages, earnings and non-labor income such as: financial income, immovable income, transfer income and other types of income.

The 'German Socio-Economic Panel (GSOEP)' is surveyed by the German Institute for Economic Research (DIW). GSOEP was launched in 1984 with 5,921 households and 12,290 individuals aged over 16. The data consist of 8 samples: *Sample A* to *Sample H*. *Sample A* covers 'Residents in West Germany (the Federal Republic of Germany).' Sample B covers 'Foreigners in West Germany,' including persons in private households with a Turkish, Greek, Yugoslavian, Spanish or Italian household head. Sample C, 'German Residents in East Germany (the German Democratic Republic (GDR)),' covers persons in private households where the household head was a GDR citizen. Sample D for 'Immigrants' was launched in 1994/95. Sample E 'Refreshment' was selected independently of the ongoing panel (subsamples A through D) from the population of private households in Germany. Furthermore, since 2002, there is also Sample F 'Innovation' and Sample G 'Oversampling of High Income.' In 2006, Sample H 'Refreshment sample' was initiated with 1,506 households selected from the total population. The data includes eight main areas in Germany and contains a large number of socio-economic variables: Population and demography, education, training, qualification, labor market and occupational dynamics, earnings, income, social security, housing, health, household production, basic orientation and satisfaction with life in general, and certain aspects of life etc. The data used for this study are taken from GSOEP Sample A to F and H between 2002 and 2007. Only those individuals with a West German educational background are selected in the analysis, which should thereby establish a comparable estimation of the different earnings effects of school education and job training between Korea and Germany.

2.4.2. Descriptive Statistics

The main variables this study uses are earnings as a dependent variable and education and job training as independent variables. I further control for age and its square, tenure and its square, marital status, household head, having children, residence of family, working in the public sector, firm size, industry, working part time, working overtime and use year dummies. This study utilizes 7,359 observations for 2,709 individuals from KLIPS and 13,011 observations for 4,129 individuals from GSOEP while employing these variables.

Labor Earnings This study uses the gross yearly labor earnings of individuals as the dependent variable. KLIPS has primarily reported the net current mean wages and salaries as its main individual income variable. From the 2003 data, KLIPS additionally offered an individual income variable, the yearly gross/net labor earnings for the last calendar year under consideration, which gave the total labor earnings level for all employees. To estimate the effects of school education and job training, the yearly labor earnings have a distinct advantage when compared with current mean wages and salaries from principal employment: Labor earnings certainly reduce sample selection bias, because this variable reports not only wages and salaries from the main job, but also wages and salaries from any secondary job etc., and this includes not only labor earnings for wage earners, but also for the self-employed. In GSOEP, the counterpart is the yearly gross labor earning in the last calendar year. Both labor earnings include wages and salary from principal jobs, those earned during job training, bonuses, allowances from overtime and profit-sharing, and earnings from the self-employment etc. In the estimation, the earnings in the year under consideration are used: Earnings data, surveyed in 2007, are combined with other variables, surveyed in 2006. Yearly mean earnings in the year under consideration are about 21,000 Euro in Korea and 38,500 Euro in Germany.

Education Level This study aims to compare earnings differentials between various educational degrees, especially between senior high school and university graduation in Korea with the comparable German educational degrees. As we have seen, school education systems in both countries are quite different. This study classifies different educational categories in both countries into equivalent educational degrees by referring to the education indicator of the UNESCO – International Standard Classification of Education 1997. The educational levels are classified into 5 levels (see Table 2.3 and Table 2.4).

Level 1 is equivalent to an education level lower than junior high school graduation or graduation from under-performing schools in Korea (i.e. 3.25% of observations). Level 1 comprises those individuals who leave school without having attained a general elementary schooling degree (under *Haupt/Realchulabschluss*) in Germany (1.18% of observations). These belong to ISCED-97 level 1. Graduates from junior high school in Korea (8.17%) and individuals completing a general elementary education (*Haupt/Realchulabschluss*) in Germany (14.48%) were classified in Level 2 corresponding to ISCED-97 level 2.

Level 3 signifies a completed senior high school education. This includes an educational qualification sufficient for university entrance – completed academic senior high school in Korea and *Abitur* in Germany –, a degree with vocational education – completed vocational senior high school in Korea and middle vocational qualification (*Beruflicher Abschluss*) in Germany, or a degree with middle vocational qualification taken after completing the *Abitur* in Germany. Thus, the education levels between the upper secondary and the lower tertiary levels are classified in this level. 42.51% of observations in Korea and 54.45% in Germany belong to level 3.

Level 4 consists of completed junior college education in Korea and higher vocational qualifications such as completing 2 years of healthcare school (*Schule Gesundheitswesen*), 2 years of technical school (*Fachschule*), or a civil servants apprenticeship (*Beamtenausbildung*) in Germany. 15.14% of observations in Korea and 10% of observations in Germany appertain to level 4. Level 5 is made up of those holding a bachelor's degree or higher from a Korean university and individuals possessing a diploma from a university of applied science (*Fachhochschuldiplom*) or higher in Germany. 30.94% observed data in Korea and 19.89% observed data in Germany relate to level 5. The Educational indicator of the ISCED-97 does not classify these levels of education – level 4 and level 5 – corresponding to educational degree. The ISCED-97 divided levels 4 and 5 according to educational context and the aim of each educational institution. But this study is interested in the earnings effects of educational degrees, especially the earnings differential between a senior high school and a university degree. So, unlike the ISCED-97 classification, this study classifies these levels of education according to the fact as to whether someone possesses at least a bachelor's degree or not.

Education Year In spite of a careful classification of educational degree, a noticeable qualitative difference in school education between both countries becomes apparent in this study. A comparable level of education differs not only in educational program, but also in educational duration. Seeking to acquire a more comprehensive picture of educational effects on earnings, this study additionally generates dummy variables using the cumulated educational years required for the completion of an educational degree. This duration is not made up of the actual years of schooling for each individual, but a standardized duration for an educational program.[18] Employing this variable for education years, I take

18 See Tables 2.5 and 2.6 for further details on the computation method used to analyze educational durations.

7,359 observations for the Korean dataset and 12,934 observations for the German data. Durations can be classified into 16 year groups in Germany and 7 year groups in Korea. I could find 5 comparable year groups from both datasets: i.e. 9 years (8.17% in Korea, 6.84% in Germany for each observation), 12 years (42.51%, 7.55%), 14years (15.14%, 1.88%), 16 years (26.77%, 3.65%) and 18 years and more (4.17%, 10.78%).

Job Training In order to estimate any job training effect on wages, this study uses the completed job training in the last period (approximately the last 12 months) for the Korean dataset and the completed job training during the last calendar year for the German dataset. The question concerning job training in KLIPS is: "*Have you done any vocational education or training to get a job, start your business or enhance productivity after the last interview?*" The comparable job training variable in GSOEP can be determined in the following way: By posing a question such as: "*Did you finish schooling, vocational training, or university education (in the last calendar year)?*" Further, a subordinate question relating to "*type of education or training*" could also be posed. The types of vocational education and training are: Apprenticeship (*Lehre*), full-time vocational school (*Berufsfachschule, Handelsschule*), school for health care professions (*Schule Gesundheitswesen*), trade and technical school for vocational education ("*Meister-, Technikerschule*" etc.), training for public employees (*Beamtenausbildung*), company retraining, further training and others. From these, apprenticeship, company retraining, further training and others are comparable to the Korean situation in job training, and are thus included in generating the German job training variable. Employing this generation method of variable and limiting observations to males between the ages of 18 and 55, the proportion of individual observations in job training increases in KLIPS and decreases in GSOEP.[19] Figures for completed job training are 15.37% in Korea and 1.58% in Germany for each observation.

Further, this study uses interaction variables relating to completed job training and education levels to estimate the earnings differentials attributable to job training for each educational level. An additional variable, job training undertaken in earlier years will show us the time lagged earnings effect of earlier job training. Descriptive statistics of other variables are presented in Table 2.7 and Table 2.8 in the appendix.

19 Exclusion of advanced job training at school and including only completed job training mainly reduces proportion of job training from the observations in Germany. Analysing only male adults increases proportion of job training from the observations in Korea.

2.5. Estimation Results

2.5.1. Estimation Results using the Variables: Education Levels and Job Training

Table 2.9 and Figure 2.1 present the effects of education levels and completed job training on earnings in Korea by means of OLS, quantile regression and the fixed effects methods. The base variable of the education level dummies is completed senior high school and the base variable of job training is no job training. Employing the quantile regression method, this study presents earnings differentials between and within education levels and job training. In general, education levels are positively correlated with greater earnings in all models. Completed job training was also found to have a positive influence on earnings. The estimation results are in most cases statistically highly significant.

According to the estimation results obtained by means of the OLS method, people with an educational level lower than junior high school earn, with all other factors fixed, about 26% less than those who completed senior high school in Korea. Estimating at the median by means of quantile regression, they are found to earn about 23% less. The earnings effects varies between -23% and -33% along with different percentile points. At the 10th quantile, the earnings differential between people with a sub-junior high school education and those who completed senior high school is relatively large: i.e. -33%. Junior high school graduates earn about 14% less at the mean as determined by the OLS method and 15% less at the median when compared with senior high school graduates. This varies between -12% and -16% at different percentile points.

Junior college graduates earn about 15% more by OLS and 12% more at the median than the base variable. The highest earnings differential is found at the 10th quantile (20%) and the lowest differential at the 90th quantile (8%). People with a university degree and higher earn 34% more than senior high school graduates according to the OLS method and 30% more at the median. At the 10th quantile, this differential increases to 40%. People who completed their job training during the last period under consideration earn 6.5% more by OLS and 6% more at the median than people without job training. The earnings differential due to job training reaches 10% at the 10th quantile and 3.7% at the 90th quantile. In general, earnings differentials within an educational level are stronger at the end of the lower quantile. This trend is identical for job training.

Figure 2.1: *Effects of Educational Levels and Job Training on Earnings in Korea*

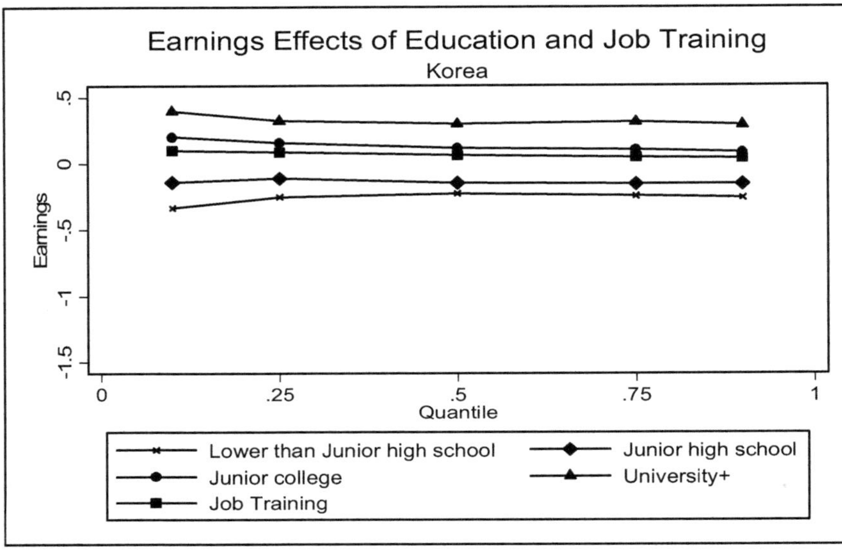

Note *Base variable for education level: Senior high school. Base variable for job training: Absence of job training*

The fixed effects method demonstrates that adult males who acquire a junior college degree have an 18% higher earnings differential when compared with males who only acquire a senior high school degree. This earnings premium is also found to be 19% if people acquire a university degree. The fixed effects method demonstrates an insignificant result for any job training effect on earnings.

Table 2.10 and Figure 2.2 illustrate earnings effects due to education level and completed job training in Germany. In most cases, these estimates are highly significant and positively correlated with education level and job training. The base variable for education is an occupational qualification (*Berufsabschluss*) and/or a qualification suitable for university admission (*Fach-/Hochschulreife*) while the base variable for job training is no job training. People with a lower than general elementary level of education (i.e. education level lower than *Haupt-/Realschulabschluss*) earn 38% less in earnings than people who have an occupational qualification and/or qualifications for university according to the OLS method. They earn 26% less at the median when compared to the base variable. The earnings differentials strongly vary along different quantile points: i.e. about -71% at the 10th quantile and -14% at the 75th quantile. Those with a general elementary degree (*Haupt-/Realschulabschluss*) earn 19% less than the base group by OLS and 15% less at the median. Those who have higher occupational qualifications, for example *Meister* and technicians, earn about 15% more as determined by OLS and 14% more at the median than the base variable. This effect is 12% at the 25th quantile and 16% at the 90th quantile. University graduates, who possess *Fachhochschuldiplom* and higher, earn 52% more according to OLS and 51% more at the median compared with the base variable. People who completed job training earn 17% more than people without job training by means of the OLS and at the median. Intra-job training earnings differentials vary from 9% to 17%. The fixed effects method, which utilizes the transformed function, illustrates 49% smaller earnings effect for people with lower than general elementary education and 41% smaller earnings effect for people with a general elementary degree when compared with the base group. Those who completed high vocational qualifications such as *Meister* and technicians have 14% higher earnings effects than those who acquire an occupational qualification and/or a qualification for university, up to 10% significance level. The earnings premium increases to 140% with the attainment of a university degree when compared to the base group. The fixed effects technique demonstrates an insignificant result for job training effect on earnings.

Comparing estimation results in Korea and Germany, earnings differentials between education levels are in general larger in Germany than in Korea, as we have seen. The earnings differential between university and senior high school graduates is also larger in Germany (about 50% using OLS and at the median) than in Korea (about 30%), in contradistinction to the hypothesis of this study. The effect of completed job training is higher in Germany than in Korea, as this study expected. This leads to about 17% earnings premium in Germany and about 6% premium in Korea using OLS and at the median. Earnings differentials are high at the lowest education level in both countries, but much higher in Germany. *Meister* and technicians in Germany and university graduates in both countries enjoy relatively even intra-educational level earnings differentials. By the fixed effects method, which utilizes the transformed function, this study reveals differences in earning effects attributable to adult education in both countries: Earnings effects for lower levels of education have a highly significant value in Germany, while these are insignificant in Korea. Attainment of a university degree has a strong effect on earnings in Germany (140%), while this is only about 19% in Korea estimated using the fixed effects method. As can be seen, attainment of university degree in Germany is much more profitable than is the case in Korea.

Figure 2.2: Effects of Educational Levels and Job Training on Earnings in Germany

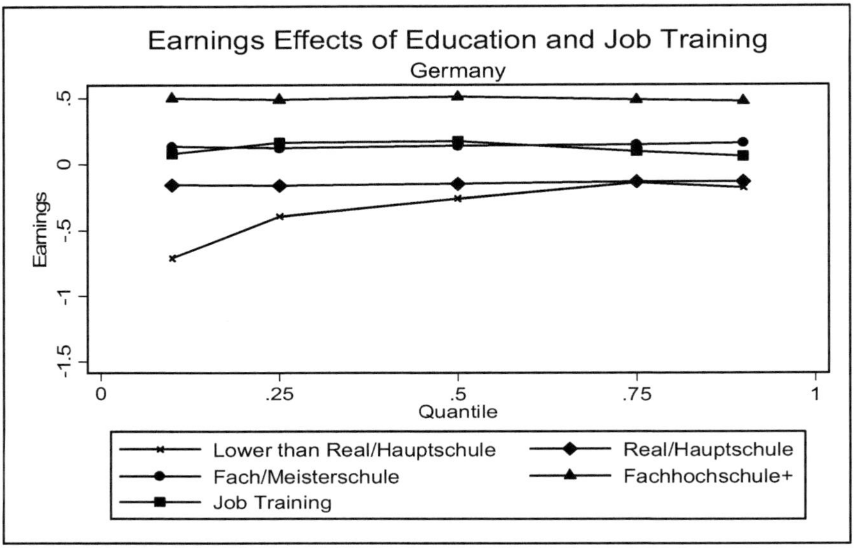

Note *Base variable for education level: Berufsabschluss and/or (Fach-)Hochschulreife. Base variable for job training: Absence of job training.*

2.5.2. Estimation Results using the Variables: Educational Duration and Job Training

Tables 2.11 and 2.12 illustrate the effects that the duration of education and job training have on present earnings in Korea and Germany. Using variables denoting years spent in education, this study demonstrates the earnings effects of education in Korea and Germany in the same quantitative categories. The base variable for the educational year dummies is 12 years education. This represents a completed senior high school education in Korea, and the completion of specialized vocational high school (2 years *Fachoberschule*) and specialized vocational high school for qualification of university entrance or occupation (*Berufsfachschule, die eine Studienberechtigung or Berufsqualifikation vermittelt*) in Germany (OECD (1999)). The estimation results indicate that more years spent in education correlate with higher earnings premiums in Korea. But this is not evident in Germany: The earnings effect of 9 years education is higher than 10 years education at the median, 11years education has a greater effect than 13 years education and 14 years education a greater effect than 14.5 years at the mean and the median etc. The lowest educational duration is represented as zero years in Korea and 7 years in Germany. By using OLS and studying at the median, zero years of education results in a 46% drop in earnings compared to 12 years of education in Korea. A value of -57% is determined at the 75th quantile and -43% at the 10th quantile. 7 years education in Germany leads to a -46% earnings differential compared to 12 years education by use of OLS and -33% at the median. The highest differential of -78% is found at the 10th quantile and the lowest differential of -27% at the 90th quantile. The intra-earnings differential of the lowest educational year group is much larger in Germany compared to Korea. Earnings differentials for those who have completed 9 years of education are about -14% in Korea and -32% in Germany by means of the OLS. At the median, these are about -15% in Korea and -28% in Germany. This varies between -12% and -16% in Korea and between -26% and -38% in Germany at the different percentile points. Between and within earnings differentials of this group are stronger in Germany than in Korea. For 14 years of education, earnings differentials are 14%-15% higher in both countries by OLS compared to 12 years education. These are 11% in Korea and 16% in Germany at the median. These vary from 8% at the 90th quantile to 20% at the 10th quantile in Korea, and from 14% at 75th quantile to 21% at the 25th quantile in Germany.

16 years of education indicates a bachelor's degree in Korea. In Germany, this duration represents an educational degree obtained from the dual system after passing the *Abitur* and a degree obtained from the upper secondary evening schools in the ISCED-level 4B, or a degree obtained from 3 years vocational education at trade and technical schools for meister and technician, 3 years at health sector schools and colleges of public administration in the ISCED-level 5B education (OECD (1999)). People with 16 years of education earn 30% more in Korea and 27% more in Germany according to the OLS method compared to people with 12 years of education. The effects are 26% in Korea and 30% in Germany at the median. It varies between 26% at the median and 39% at the 10th quantile in Korea by means of quantile regression. In Germany, these figures are between 22% at the 90th quantile and 32% at the 25th quantile. The difference in earnings between senior high school (12 years of education) and a bachelor's degree at the university (16 years of education) in Korea ranges approximately similar to its German counterpart when measured in years of education. 18 years plus of education are equivalent to the educational level attained via a diploma, magister and doctorial degree at university in Germany and master and doctoral degree at university in Korea. The earnings premium in this group reaches 60% in Korea and 55% in Germany (OLS) and 54% in Korea and 51% in Germany at the median. This equals 70% in Korea and 60% in Germany both at the 10th quantile. For this group, inter and intra-earnings differentials are slightly higher in Korea compared with in Germany (see Figures 2.3 and 2.4). By means of the

fixed effects method, 16 years and 18 years+ of education lead to very high increases in earnings with values of 2.97 and 7.11 in Germany compared to 12 years of education. These effects are much higher in Germany than in Korea (values of 0.19 and 0.24). Fewer years of education have a highly significant and strongly negative effect on earnings in Germany, while fewer years yield insignificant results in Korea using the fixed effects method.

Completed job training leads to a 6% earnings increase in Korea and a 10% increase in Germany, when compared to no job training (OLS). At the median, this leads to a 7% earnings increase in Korea and an 11% increase in Germany. Job training effects on earnings are higher in Germany than in Korea. However, job training effects become lower and often yield insignificant results in Germany compared to the estimation model where educational degree is the prime factor as opposed to educational duration. This might be the result of a relatively strong correlation between years of education and completed job training rather than any correlation with educational degree.

2.5.3. Job Training Effects on Earnings including Interaction Terms and Variables for Earlier Years in Job Training

Table 2.13 illustrates the effect of education and job training in Korea including interaction variables for education levels and completed job training. These interaction terms demonstrate the effects of completed job training on earnings between educational levels. The interaction effect of education level and completed job training is in general higher for lower education levels and lower for high education levels. Completed job training for an education level below junior high school graduation leads to a 26% higher earnings effect than the base group – the completed job training in the education level of senior high school – by means of OLS. The effect of completed job training at an educational level equivalent to junior high school is 18% higher than the base variable at the median. This reaches 39% at the 75th quantile. Completed job training for junior college graduation results in a 10% decrease in the earnings effect when compared with the base variable using OLS. For university graduation, this results in a 5% drop in the earnings effect (OLS), and an 8% drop at the median. This attains the value of -14% at the 10th quantile.

Table 2.14 illustrates the effects of education, job training and the interaction terms in Germany. There is a clear trend for the interaction effect to become higher at a lower educational level – *Haupt-/Realschulabschluss* – and much lower at higher educational levels. Completed high vocational qualifications at school such as *Fach-/Meisterschule* results in a decrease of 21% in the earnings effect compared to the base variable – the group of occupationally qualified and/or the qualified group for university – at the median. This attains to -37% at the 10th quantiles. For university graduation, these figures are -38% utilizing OLS and -21% at the median. At the 10th quantile, this negative effect reaches by -67%. Compared with Korea, these job training effects for highly educated people are many folds smaller in Germany. Job training compensates for any education-earnings inequality, especially in Germany.

According to Becker's human capital theory (1962), human capital investment could have an effect on earnings in the long run. This study therefore includes any completed job training undertaken in earlier years to identify levels of job training effect on earnings more definitely. From the German dataset, this study can estimate that the earnings effects of completed job training in the past 7 years and from the Korean dataset these effects in the past 5 years. After controlling for the effect of earlier job training, job training for the previous year in Germany does not present a significant effect any longer, while completed job training for the last period in Korea still gives a significant result. Completed job training undertaken 3, 4, 5 and 7 years ago has a significant earnings effect in Germany: 15%, 13%, 13-17% up to the usual significance level. In Korea, completed job training undertaken in the last period, 5 years ago and earlier results in a 6% to 8% higher earnings differential. The job training effect on earnings demonstrates a significant time lag in Germany. The level of earnings effects including variables of any completed job training undertaken in earlier years demonstrates approximately similar levels of earnings effects, which are estimated only using variable of completed job training in the last year/period in both countries (see Table 2.15).

2.5.4. Other Findings

This study further finds earnings effects related to age and tenure along with some family and firm's characteristics. Concerning ***age*** and ***tenure***, age-earnings profiles have a greater effect on earnings in Germany than in Korea. Tenure-earnings profiles indicate a much stronger effect on earnings in Korea than in Germany in all models. The earning effect of age attains its highest value between the ages of 42 and 43 in Korea, and 44 and 45 in Germany at the mean and median. The tenure effect on earnings is at its greatest between the tenure of 31 and 32 at the mean, and 39 and 40 at the median in Korea. The greatest tenure effect arises between the tenure of 26 and 27 at the mean, and 32 and 33 at the median in Germany (see Tables 2.16 and 2.17, Figures 2.5 to 2.8).

Concerning ***marital status***, married couples earn 8% more in Korea and 12% more in Germany according to OLS. At the median, married couples earn 6% more in Korea and 7% more in Germany compared to singles. Within earnings differentials for married couples are greater in the lower quantiles – 10th and 25th quantile – for both countries. Household heads earn 10% and 9% more in Korea, and 3% and 3.5% more in Germany (OLS) and at the median compared to other household members. Earning differentials due to children are higher in Korea (3% to 7%) than in Germany (2% to 3%) at the different quantiles (see Tables 2.18 and 2.19, Figures 2.9 to 2.11).

No significant results are obvious in the ***public sector*** in Korea, whereas highly significant results are evident in Germany. These are equal to -5% (OLS), -7% at the median, -12% at the 75th quantile and -18% at the 90th quantile in Germany. At the higher quantiles, public sector employees suffer a greater loss in earnings when compared to private sector employees. ***Working overtime*** generally leads to significant positive earnings effects in both countries. These effects are higher in Germany than in Korea: 14.6% in Germany and 9.3% in Korea (OLS), and 11% in Germany and 9% in Korea at the median. At the 10th quantile, these result in the highest values and at higher quantile the lowest values in both countries. ***Part time working*** leads to a highly significant and strongly negative earnings effect for both countries. At the mean, part time working has a 35% negative earnings effect in Korea and a 32% negative effect in Germany, compared to full time working. These differentials are apparent at the 10th quantile: -58% in Korea and -51% in Germany. The lowest values are found at the 90th quantile: -15.5% in Korea and -18.5% in Germany. By utilizing the fixed effects method we find that part time working has a -18% earnings effect in Korea and -2% in Germany (see Tables 2.20 and 2.21, Figures 2.12 to 2.14).

Concerning ***firm size***, this study found the classification of similarly sized firms into the same categories from different datasets: KLIPS and GSOEP to be particularly difficult. This study could generate the same base variable - firms with 1 to 4 employees -, and approximately divide the firm size into 5 similar categories. Most estimation results are highly significant and have a positive effect on earnings, compared to the base groups in Korea and Germany. A larger firm size leads to a stronger earnings effect. Earnings effects related to the size of the firm are stronger at the lower quantile and relatively weak at the higher quantile in both countries. Intra-firm size earnings differentials are larger in Germany than in Korea (see Tables 2.20 and 2.21, Figures 2.15 and 2.16).

Base variables of ***family residence*** are the capital city, Seoul, in Korea and the western part of the capital city, West Berlin, in Germany. So, this study could indicate earnings differentials between those who live in the capital cities and those who live in other regions. People who live in Seoul have higher earnings compared with people who live in all other regions of Korea. People who live in West Berlin, however, only earn more when compared to people who live in the eastern part of Germany. At the 90th quantile, people who live in other regions of Germany earn less than people who live in West Berlin. The estimation results reveal the significance of the earnings inequality problem between people living in the capital city, Seoul, and in other Korean regions compared with in Germany (see Table 2.22 and Table 2.23).

2.6. Conclusion

This study questions why the possession of a university degree is considered to be so much more important in Korea than in Germany. The hypothesis is that the earnings gap between university graduates and senior high school graduates or those with a lower level of education is much higher in Korea than in Germany. A major reason is considered to be that well developed German vocational education at school as well as job training for adults reduces earnings gaps between these education levels. This lessens competition for university entrance in Germany. This study estimates earnings differentials between education levels and job training for both countries by employing the quantile regression method together with the OLS and fixed effects methods.

Educational Degree Comparing Korea and Germany, this study found higher earnings differentials due to education levels in Germany than in Korea. By using OLS, the differences between the lowest education level and senior high school level education are about -38% in Germany and about -26% in Korea. At the median, these differences are about -26% in Germany and about -23% in Korea. Earnings differences between senior high school and the highest educational group (i.e. university graduates) are also higher in Germany than in Korea: The percentages are 34% (OLS) and 30% at the median in Korea, and 52% (OLS) and 51% at the median in Germany. Well developed German vocational education at school does not seem to reduce earnings differentials between educational degrees as strongly as this study expected. Intra-educational earnings differentials are relatively high for the lowest educational level when compared to other levels, and relatively low for university graduates in both countries. Especially in Germany, the earnings effect varies widely for those at the lowest educational level.

Educational Year Although earnings are apparently positively correlated with years of education in Korea, a longer educational duration does not always assure the higher earnings premium found in Germany. Inter and intra-educational earnings differentials associated with 14 years and 16 years of education are approximately similar in both countries. For 18 years and longer education, these differentials are slightly higher in Korea. People with 16 years of education earn 31% more in Korea and 27% more in Germany compared to people with 12 years of education according to the OLS (26% in Korea and 30% in Germany at the median): i.e. the earnings differential between a bachelor's degree (16 years education) and senior high school graduation (12 years education) in Korea does not result in any apparent difference in earnings when compared to the same number of years of study in Germany. This similarity, however, does not support the hypothesis of this study: Neither the level of education, nor its duration demonstrates a higher inter-educational earnings differential in Korea than is the case in Germany.

By means of the fixed effects method, earnings effects attributable to a lower educational level are highly significant in Germany, while they are insignificant in Korea. This may demonstrate the inflexibility of education in relation to age: Koreans who could not go to school at an expected time can hardly fill this educational gap as adults.

Job Training In the model with educational degree, job training results in about 17% higher earnings differential using OLS and at the median in Germany, while this percentage is 6% in Korea. Using educational year variables, job training effects are determined to be 10% and 11% in Germany and 6% and 7% in Korea respectively. The job training effect on earnings is stronger in Germany than in Korea, as expected. Concerning the absolute dimension of job training effects, however, we should bear in mind that these estimates suffer from selection bias and omitted ability bias. The real effect is usually smaller than estimates derived from this study. Using the interaction terms of completed job training and education levels, the job training effect on earnings is found to be higher at a lower educational level and lower at a higher educational level in both countries. This trend is much more apparent in Germany than in Korea: -38% (OLS) and -21% (median) for university graduates in Germany, and -5% (OLS) and -8% (median) for university graduates in Korea. Job training more strongly reduces earnings differences between educational levels in Germany than in Korea, as expected.

Other Findings Age effects on earnings are stronger in Germany, but tenure has stronger effects on earnings in Korea. The earning differential between married couples and singles is stronger in Germany. Maybe, the level of earnings influences the possibility of marriage in Germany unlike Korea, where age is a more decisive factor for marriage. The earnings difference between a household head and other family members is greater in Korea, and having children also has a higher earnings effect in Korea. Earnings structures are more family oriented in Korea. People who live in the capital city, Seoul, earn more than people who live in other regions in Korea. However, people who live in West Berlin only earn more than people who live in East Germany. This result reveals a serious earnings/economic inequality between the capital city, Seoul, and other regions in Korea. Working overtime is more rewarding in Germany than in Korea. Part time working leads to a highly significant and strongly negative earnings effect (about -32% and -35% by OLS) for both countries. This difference must be hard for a male adult, unwillingly working on a part time basis.

Earnings differences attributable to education levels are lower in Korea than in Germany, the opposite of the hypothesis of this study. The earnings differential between senior high school and university graduates is also apparently lower in Korea than is the case in Germany. Job training effect is higher in Germany than in Korea. Job training offers people at a lower educational level a better opportunity to enhance their earnings in Germany. Job training has a much lower effect on earnings for German university graduates compared to low educated people. However, this effect does not seem to be strong enough to reduce earnings differentials between educational levels. The well-developed system of vocational education in Germany evidently does not compensate for the earnings differentials between university graduates and people with less education. The estimation results do not support the explanation: Higher return due to university education in Korea increases the desire to go to university compared with Germany. The high earnings differential between university graduates and people at a lower educational level in Germany may be attributable to the longer duration and/or the higher productivity of university education. And/or the lower number of university graduates in Germany leading to their higher earnings in the labor market. This speculation could be an interesting topic for future research.

Why is a university education in Korea held to be so much more important than in Germany? Koreans need to find an alternative answer to this question. The introduction of the German system in Korea would certainly offer a chance to enhance vocational education, but this enhancement would not automatically reduce earnings differences between educational levels or lessen the excessive competition for university entrance in Korea.

References

Ashenfelter, O. and Krueger, A. (1992), "Estimates of the Economic Return to Schooling from a New Sample of Twins," NBER Working Papers, 4143, National Bureau of Economic Research.

Ashenfelter, O. and Zimmerman, D.J. (1993), "Estimates of the Returns to Schooling from Sibling Data: Fathers, Sons, And Brothers," NBER Working Papers, 4491, National Bureau of Economic Research.

Ashenfelter, O. and Zimmerman, D.J. (1997), "Estimates of the Returns to Schooling from Sibling Data: Fathers, Sons, And Brothers," *Review of Economics and Statistics*, 79(1), pp. 1-9.

Becker, G.S. (1962), "Investment in Human Capital: A Theoretical Analysis," *Journal of Political Economy*, LXX, pp. 9-49.

Behrman, J.R., Hrubec, Z., Taubman P. and Wales, T.J. (1980), *Socioeconomic Success: A Study of the Effects of Genetic Endowments. Family Environment and Schooling*, Amsterdam: North-Holland.

Beicht, U., Krekel, E.M. and Walden, G. (2006), *Continuing Vocational Education and Training - What Costs, What Benefits Do Participants Have?*, Bielefeld: Bertelsmann.

Blackburn, M.L. and Neumark, D. (1991), "Omitted-Ability Bias and the Increase in the Return to Schooling," NBER Working Papers, 3693, National Bureau of Economic Research.

Bundesinstitut für Berufsbildung (2006), *Neue und modernisierte Ausbildungsberufe 2006: Kurzbeschreibungen.* Bonn.

Bundesministerium für Bildung, Wissenschaft, Forschung und Technologie (1997), *Reformprojekt Berufliche Bildung – Flexible Strukturen und moderne Berufe*, Bonn.

Center for Educational Statistics (2009), *Educational Statistics*, http://cesi.kedi.re.kr /index.jsp, (2009.12.15), Seoul (in Korean).

Garen, J. (1984), "The Returns to Schooling: A Selectivity Bias Approach with a Continuous Choice Variable," *Econometrica*, 52(5), pp. 1199-1218.

German Institute of Economic Research (2009), *German Socio Economic Panel – Documentation PGEN*, www.diw.de/documents/dokumentenarchiv/17/diw_01.c.60055.de/pgen.pdf, (2009.11.30), Berlin.

Griliches, Z. (1979), "Sibling Models and Data in Economics: Beginnings of a Survey," *Journal of Political Economy*, 87(5), pp. 37-64.

Hámori, S. (2007), *Essays in Labour Economics and Economics of Education*, Dissertation, Universität Mannheim.

Heckman, J.J. (1979), "Sample Selection Bias as a Specification Error," *Econometrica*, 47(1), pp. 153-161.

Heckman, J.J. and Robb, R. (1985), "Alternative Methods for Evaluating the Impact of Interventions," in: Heckman, J.J. and Singer, B. (Eds.), *Longitudinal Analysis of Labor Market Data*, Econometric Society Monographs Series, 10, Cambridge: Cambridge University Press, pp. 156-245.

Hippach-Schneider, U., Krause, M. and Woll, C. (2007), *Berufsbildung in Deutschland – Kurzbeschreibung*, CEDEFOP (European Centre for the Development of Vocational Training) Panorama Series, 136, Luxemburg: Amt für amtliche Veröffentlichungen der Europäischen Gemeinschaften.

Kim, A.C., Park, J.Y., Kim, H.J., Yoon, J.H. and Kim, T.E. (2006), *The Reorganization Program of the School System for the Future*, Seoul: Korean Educational Development Institute (in Korean).

Kim, K.H. (1998), "Die neue Entwicklungstendenz der Berufsbildung in Deutschland und Hinweise auf unsere Berufsbildungspolitik," *Koreanisch-Deutsche Forschung der Erziehungswissenschaft*, 3(1), pp. 27-38 (in Korean).

Kim, K.H. (2006), "The Vocational Education System and the Priority given to Talent and Attitude in Education," *Eduforum*, 124, pp. 17-19 (in Korean).

Koenker, R. (2005), *Quantile Regression*, Cambridge: Cambridge University Press.

Koenker, R. and Bassett, G. (1978), "Regression Quantiles," *Econometrica*, 46 (1), pp. 33-50.

Koenker, R. and Hallock, K.F. (2001), "Quantile Regression," *Journal of Economic Perspectives*, 15(4), pp. 143–156.

Kolja Briedis (2009), "Wo stehen Hochschulabsolventinnen zehn Jahre nach dem Abschluss?", *HIS Magazine*, 3/2009, Hannover: Hochschul-Informations-Systems.

Korean Educational Development Institute (2005), *Educational Indicators for Primary and Secondary Levels of Schooling*, RM2005-56, Seoul (in Korean).

Korea Labor Institute (2008), *Korean Labor & Income Panel Study,* http://www.kli.re.kr/klips/ko/main/main.jsp, (2010.02.15), Seoul (in Korean).

Korean Statistical Information Service (2010), *The Number of Unemployed People according to Age and Education Level*, http://kosis.kr/nsportal/themes/themes _04List.jsp, (2010.06.12), Seoul (in Korean).

Kuwam, K., Bilger, F., Ganhs, D. and Seidel, S. (2006), *Berichtssystem Weiterbildung IX: Integrierter Gesamtbericht zur Weiterbildungssituation in Deutschland*, Bonn, Berlin: Bundesministerium für Bildung und Forschung.

Lee, B.H., Kim, J.S., Kim, T.J., Jang, S.M. and Jang, W.S. (2006), *A Study of Life Long Education and its Effects on the Labor Market*, Policy Research, 2006-08, Seoul: Korea Labor Institute (in Korean).

Funk, L. (2005), "Situation and Vision of the Vocational Educational Training System in Germany," *International Labor Brief*, 3(6), pp. 47-57 (in Korean).

Machado, J.A.F. and Mata, J. (2001), "Earnings Functions in Portugal 1982-1994: Evidence from Quantile Regressions," *Empirical Economics*, 26(1), pp. 114-134.

Mincer, J. (1974), *Schooling, Experience and Earnings*, New York: Columbia University Press.

Ministry of Education, Science and Technology (1997), *Curricula for the Upper Secondary School Education 2*, announcement no. 1997-15, Seoul (in Korean).

Ministry of Education, Science and Technology (2009), "Selecting Meister-schools," Press Release 2009.02.12, Seoul (in Korean).

Münch, J. (1994), *Das Berufsbildungssystem in der Bundesrepublik Deutschland*, European Centre for the Development of Vocational Training (CEDEFOP), Berlin: Amt für amtliche Veröffentlichungen der Europäischen Gemeinschaften.

OECD (1999), *Classifying Educational Programmes – Manual for ISCED-97 Implementation in OECD Countries*, Paris.

OECD (2007), *Education at a Glance 2007 – OECD Indicators*, Paris.

OECD (2008), *Education at a Glance 2008 – OECD Indicators*, Paris.

Park, Y.G. (2007), "Human Capital Development for Industrialization," http://contents.archives.go.kr, (2009.12.05), Seoul: National Archives & Records Services (in Korean).

Reuling, J. and Hanf, G. (2003), *The Role of National Qualifications Systems in Promoting Lifelong Learning - Background Report for Germany*, Bonn: Bundesinstitut für Berufsbildung.

Suh, S.S. (2002), *Course of Vocational Training in Korea: The Back Story of Institutionalization*, Seoul: Korea Chamber of Commerce and Industry (in Korean).

Appendix

Table 2.3: Classification of Education Levels

Classification	ISCED	Korea	Germany
Level 1 Less than junior high school level of education	-	Lower than elementary schooling or inadequate	Inadequate *(Anderer Abschluss)*
	ISCED 1 Primary level of Education	Elementary school	No degree *(ohne Abschluss verlassen)*
Level 2 Completed junior high school level of education	**ISCED 2** Lower secondary level of education	Junior high school	General elementary *(Haupt/ Realchulabschluss)*
Level 3 Completed senior high school level education: Upper secondary and lower tertiary level of education	**ISCED 3** Upper secondary level of education	Senior high school	Qualification for college or university or middle vocational qualification *(Fach-/Abitur or Beruf. Abschluss)*
	ISCED 4 Post-secondary and non-tertiary		Qualification for college or university and middle vocational qualification *(Beruf. Abschluss nach absolv. allg. Schule)*
Level 4 Junior college or higher vocational education	**ISCED 5B** First stage of tertiary education	Junior college	Higher vocational qualification *(Höherer beruflicher Abschluss: Meister, Techniker)*
Level 5 University and more	**ISCED 5** First stage of tertiary education	Bachelor Master Doctor at university	Higher education: University+ *(Fach-/Hochschule or Universität)*
	ISCED 6 Second stage of tertiary education		

Sources: *OECD (1999), German Institute of Economic Research (2009) and Korea Labor Institute (2008).*

Table 2.4: Comparison of School Education in Germany and Korea according to ISCED-97

***Contents of the table**: Name of educational program (its German or Korean name)/ completion level of each program (typical starting age of relevant schooling, length of the programs and cumulative years of schooling at the end of programs)*

Primary and lower secondary level	
Germany	**Korea**
ISCED 1	
Primary school (Grundschule)/ no degree (6 years old, 4-6 program years, 4-6 total schooling years)	Civic school (공민학교)/ certification (usually adult, 3 program years, 3 total schooling years) Elementary school (초등학교)/ certification (5-6, 6, 6) Special school, primary school course (특수학교 초등과정)/ certification (6+, 6, 6)
ISCED 2A	
Lower secondary school, no access to general (Hauptschule, Realschule)/ Haupt-, Realschulabschluss (10, 6, 10) Lower secondary school, access to general (Gynasium, Integrierte Gesamtschule, Freie Waldorfschule)/ Realschulabschluss (18-35, 6, 10) Kollegschule copmleting with intermediate school certificate (Kollgschule, die mittleren Bildungsabschluss vermitteln)/ Realschulabschluss or equivalent (17-18, 1, 11) Lower secondary school evening school/ Haupt- or Realschulabschluss (18-35, 2, 12)	Miscellaneous school, middle school course (각종학교 중학교과정)/ certification (11-12, 3, 9) Civic high school (고등공민학교)/ certification (11-12, 3, 9) Middle school: Junior high school in this study (중학교)/ certification (11-12, 3, 9) Middle school attached to industrial firm (산업체 부설 중학교)/ certification (11-12, 3, 9) Special evening classes for working youth, middle school(특별학교, 중학교)/ certification (11-12, 3, 9) Special school, middle school course (특수학교 중학교과정)/ certification (12+, 3, 9)

Vocational extension school (Berufsaufbauschule)/ Realschulabschluss, Fachschulreife (18-22, 1, 14) Pre-vocational training year (Berufsvorbereitungsjahr)/ Abschlusszeugnis Berufsvorbereitungsjahr) (16-18, 1, 11)	

Source: *OECD (1999).*
Note *ISCED 2A: Programs are designed to prepare students for direct access to ISCED level 3 in a sequence which would ultimately prepare students for attending tertiary education, that is, entrance to ISCED 3A or 3B.*

Upper secondary and post-secondary non-tertiary	
Germany	**Korea**
ISCED 3C	
	Trade high school (고등기술학교)/ certification (14-15, 3, 12) Vocational high school: Vocational senior high school in this study (실업고등학교)/ certification (14-15, 3, 12) Special evening classes for working youths, vocational high school (실업계 특별학교)/ certification (14-15, 3, 12) Acrredited non-formal education facilitites school (사회교육시설학교)/ certification (14-15, 3, 12)
ISCED 3B	
Specialized vocational schools: Intermediate school certification (Berufsfachschulen, die einen mittleren Bildungsabschluss vemitteln)/ Realschulabschluss or equivalent (16-17, 1-2, 11)	

Basic vocational training year (Berufsgrundbildungsjahr)/Abschlusszeugnis für Berufsgrundbildungsjahr (16-18, 1, 11) Specialized vocational schools: Basic vocational knowledge (Berufsfachschulen, die berufliche Grund kenntnisse vermitteln)/ Abschlusszeugnis für Berufsfachschule (16-17, 1, 11) Health sector school-1year (Schulen des Gesundheitwesens) / Abschusszeugnis für medizinische Hilfsberufe (19-20, 1, 14) Kollegschule: Occupational qualification (Kollegschulen, die einen Berugsabschluß vermitteln)/ Beruflicher Abschluß (17-18, 2-3, 12) Specialized vocational schools: Occupational qualification (Berufsfachschulen, die einen Berufsabschluß vermitteln)/ Beruflicher Abschluß (16-17, 3, 13) Dual system (Duales System, Berufsschulen)/ Lehrabschluß (16-18, 3, 13)	
ISCED 3A	
Specialised vocational high schools (Fachoberschulen)/ Fachhochschulreife (16-18, 2, 12) Specialised vocational high schools: Qualification for ISCED 5A (Berufsfachschulen, die eine Studienberechtigung vermitteln)/ Fach- or Hochschulreife (16-18, 2-3, 12) Kollegschulen: Qualification for ISCED 5A, also in some cases occupational qualification (Kollegschulen, die eine Studienberechtigugn vermitteln) / Fach- or Hochschulreife and Beruflicher Abschluß (17-18, 2-3, 12)	Miscellaneous school, high school course (각종학교 고등학교과정)/ certification (14-15, 3, 12) General high school: Senior high school in this study (일반계 고등학교)/ certification (14-15, 3, 12) High school attached to industrial firms (산업체부설 고등학교)/ certification (14-15, 3, 12) Air and correspondence high school (방송통신 고등학교)/ certification (14-15, 3, 12)

Fachgymnasium/ Hochschulreife (16-17, 3, 13) Uppersecondary schools (general) / Abitur Hochschulreife (16-17, 3, 13)	Special evening classes for working youths, general high school (일반계 특별학교)/ certification (14-15, 3, 12) Special school, high school course (특수학교 – 고등학교과정)/ certification (15+, 3, 12)
ISCED 4	
ISCED 4B Dual system (Dueales System, Berufsschulen) / Lehrabschluß (19-21, 3, 16) **ISCED 4A** Specialised vocational high schools -1 year (Fachoberschulen)/ Fachhochschulreife (19-20, 1, 14) Berufsoberschulen/Technische Oberschulen/ Hochschulreife (19-20, 2, 15) Dual system (Duales System, Berufsschulen)/ Lehrabschluß (19-21, 3, 16) Upper secondary evening schools/ Abitur (Hochschulreife) (19-35, 3, 16)	

Source: *OECD (1999).*

Note *ISCED 3A: programs at Level 3 designed to provide direct access to ISCED 5A. ISCED 3B: programs at Level 3 designed to provide direct access to ISCED 5B. ISCED 3C: programs at Level 3 designed to prepare students for direct entry into the labor market, although they also provide access to ISCED 4 programs or other ISCED 3 programs. Upper secondary apprenticeship programs would fall into this category unless the programs were primarily designed to prepare students to enter ISCED 5. ISCED 4A: programs at Level 4, designed to provide direct access to ISCED 5A. ISCED 4B: programs at Level 4, designed to provide direct access to ISCED 5B.*

Tertiary education	
Germany	**Korea**
ISCED 5B	
Bavaria specialized academy (Fachakademie Byern)/ Abschluss der Fachakademie: Fachhochschulreife in some case (19-20, 2, 15)	Junior college (전문대학)/ certificiation (17-18, 2-3, 14-15)
Health sector schools-2 years (Schulen des Gesundheitswesen)/ Medizinische Assistenten (19-20, 2, 15)	Polytechnic college (기능대학)/ certification (17-18, 2, 14)
Trade and technical schools-2 years (F412)/ Fachschulabschluss, Meister,Techniker, Erzieher (21-23, 2, 15)	Misellaneous school, junior college course (각종전문대학)/ certification: some schools cannot confer certification (17-18, 2, 14)
Trade and technical schools-3years (Fachschulen)/ Fachschulabschluss, Meister, Techniker, Erzieher (21-23, 2, 16)	Technical college (기술대학)/ certification (17-18, 2-4, 14-16)
Trade and technical schools 4years (Fachschulen)/ Fachschulabschluss, Meister/Techniker, Erzieher (21-24, 4, 17)	Third military academy (육군삼사관학교)/ Bachelor (19-29, 2, 16)
Health sector schools 3 years (Schulen des Gesundheitswesen)/ Krankenschwester/-pleger (19-20, 3, 16)	National college of taxation (세무대학)/ Bachelor (17-18, 4, 16)
Schleswig-Holstein vocational academies (Berufsakademien Schleswig-Holstein)/ Berufsakademie-Abschluss (20-22, 3, 16)	Miscellaneous school, undergraduate course (각종대학)/ certification: some schools cannot confer certification (17-18, 4, 16)
Baden-Wuerttemberg vocational academies (Berufsakademien Baden-Wuerttemberg)/ Diplom(BA) (19-20, 3, 16)	Open university, polytechnic university (산업대학)/ Bachelor (17-18, 4, 16)
Colleges of public administration (Verwaltungsfachhochschulen)/ Diplom (FH) (19-20, 3, 16)	Military academy (육군사관학교)/ Bachelor (17-18, 4, 16)
	National college of police (경찰대학)/ Bachelor's degree credential for civil servant of policy (17-18, 4, 16)
	University of education (교육대학)/ Bachelor (17-18, 4, 16)
	(Military) Nursing academy

	(국군간호사관학교)/ Bachelor (17-18, 4, 16) Naval academy (해군사관학교)/ Bachelor (17-18, 4, 16) Air force academy (공군사관학교)/ Bachelor (17-18, 4, 16) School of national securities (국방대학원)/ Master (21+, 2, 18) Graduate school, special (특수대학원)/ Master (21+, 2-3, 18-19) Graduate school, professional (전문대학원)/ Master (21+, 2.5-3, 18.5)
ISCED 5A	
Fachhochschulen (Fachhochschulen)/ Diplom (FH) (19-24, 4-5, 17) Universität (University studies)/ Diplom oder Staatspruefung (19-24, 5-6.5, 18)	Air and correspondence university, open university (방송통신대학)/ Bacgelor (17-18, 2-4, 14-16) University (대학교)/ Bachelor (17-18, 4, 16) Korea advanced institute of science and technology (한국과학기술원) / Bachelor (17-18, 4, 16) The Korean National University of Arts (한국예술종합학교)/ Bachelor (17-18, 4, 16) University, medical-dentistry (의대, 치대)/ Bachelor (17-18, 6, 18) The Academy of Korean Studies, MA course (한국정신문화연구원 석사과정)/ Master (21+, 2-3, 18-19)

	Grauates school, Master's degree programme (일반대학원 석사과정)/ Master (21+, 2-3, 18) Korea advanced institute of science and technology (한국과학기술원 석사과정)/ Master (21+, 2-3, 18) University of graduate school (대학원대학 석사과정)/ Master (21+, 2-3, 18) The Korean National University of Arts, MA course (한국 예술종합학교 전문예술사과정)/ Master (21+, 2, 18)
ISCED 6	
Doctorial study (Promotionsstudium)/ Doctor(25-29, 2-4, 20)	Korea advanced institute of science and technology, PhD course (한국과학기술원 박사과정)/ Doctor (23+, 3-5, 21) Academy of Korean Studies, PhD course (한국정신문화연구원 박사과정)/ Doctor (23+, 3-5, 21) Graduates school, Doctorate program, PhD course (일반대학원 박사과정) / Doctor (23+, 3-5, 21) University of graduate school, PhD course (대학원대학 박사과정)/ Doctor (23+, 3-5, 21)

Source: *OECD (1999).*

Table 2.5: Educational Duration in Germany

Years of education = Cumulated years of schooling + Years of occupational training		Years
Schooling	No degree	7
	Lower school degree	9
	Intermediary school	10
	Degree for a professional coll.	12
	High school degree	13
	Other	10
Additional occupational training (includes universities)	Apprenticeship	1.5
	Technical schools (incl. health)	2
	Civil servants apprenticeship	1.5
	Higher technical college	3
	University degree	5

Source: *German Institute of Economic Research (2009).*

Table 2.6: Educational Duration in Korea

Korea	Cumulated educational years
No schooling	0
Elementary	6
Middle school	9
High school	12
Junior college	14
Bachelor	16
Master	18
Doctor	21+

Note *The cumulated years are standard length of schooling at the end of educational program.*

Table 2.7: Descriptive Statistics for the Korean Data KLIPS (7,359 observations for male employees aged between 18 and 55)

Variable	Mean	Std. Dev	Min	Max
Ln earnings	9.792968	0.575897	4.945613	12.83526
Yearly gross earnings in Euro	20,908.82	12,759.26	140.557	375,215.8
Age	36.94007	8.40013	18	55
Tenure	5.931116	6.487449	0	36.5
Complete education level				Percent
Edu1: Under Junior high school				3.25
Edu2: Junior high school				8.17
Edu3: Senior high school				42.51
Edu4: Junior college				15.14
Edu5: University +				30.94
Completed job training				15.37
Education*completed training				Percent
Edu1*comp. training				0.31
Edu2*comp. training				0.49
Edu3*comp. training				5.38
Edu4*comp. training				2.28
Edu5*comp. training				6.9
Household head				74.7
Having children				57.01
Marital				Percent
Single				26.63
Married				70.69
Widowed				0.26
Divorced				1.98
Separated				0.43
Public sector				5.45
Overtime				35.62
Part time				1.79
Firm size				Percent

under 5 employees				10.57
5-9				10.95
10-99				34.47
100-999				21.1
1000 and more				22.9
Residence				Percent
Seoul				21.43
Busan				9.62
Daegu				6.56
Deajeon				3.08
Incheon				6.84
Gwangju				2.98
Ulsan				4.96
Gyunggi				20.97
Gangwon				1.14
Chungbuk				2.12
Chungnam				3.02
Jeonbuk				2.54
Jeonnam				2.26
Gyungbuk				3.76
Gyungnam				8.71
Jeju				0.01
Year				Percent
2002				17.48
2003				19.08
2004				20.23
2005				21.35
2006				21.86

Note *The description of 33 industry dummies is omitted.*

Table 2.8: Descriptive Statistics for the German Data GSOEP (13,011 observations for male employees aged between 18 and 55)

Variable	Mean	Std. Dev.	Min	Max
Ln earnings	10.36009	0.724406	5.703783	12.63284
Yearly gross earnings in Euro	38,484.53	22,631.46	300	306,460.3
Age	37.93405	9.663973	18	55
Tenure	10.72565	9.482029	0	42.3
Complete education level				Percent
Edu1: Under Real/Haupt				1.18
Edu2: Real/Hauptschule				14.48
Edu3: Berufsabschluss and/or (Fach-)Hochschulreife				54.45
Edu4: Technik/ Meisterschule				10
Edu5: Fachhochschul+				19.89
Completed Job training				1.58
Education*completed training				Percent
Edu1*comp. training				0
Edu2*comp. training				0.03
Edu3*comp. training				1.41
Edu4*comp. training				0.08
Edu5*comp. training				0.05
Household head				66.26
Having children				48.1
Marital				Percent
Single				34.27
Married				58.66
Widowed				0.22
Divorced				5.3
Separated				1.55
Public sector				23.33

Overtime				56.69
Part time				11.51
Firm size				Percent
under 5 employees				4.67
5-19				13.8
20-99				18.42
100-1999				33.86
2000 and more				29.25
Residence				Percent
Berlin (West)				1.64
Schleswig-Hols				3.21
Hamburg				1.54
Lower Saxony				10.78
Bremen				0.71
N-Rhein-Westfa				26.13
Hessen				8.38
R-Pfalz,Saarl.				7.59
Baden-Wuerttemb.				15.26
Bavaria				18.36
Berlin (East)				0.44
Mecklenburg-V.				0.61
Brandenburg				1.39
Saxony-Anhalt				1.01
Thueringen				1.08
Saxony				1.87
Year				Percent
2002				22.98
2003				20.59
2004				20.44
2005				17.78
2006				18.21

Note *The description of 28 industry dummies is omitted.*

Table 2.9: Estimation Results using Variables - Educational Level and Job Training in Korea

Korea	OLS	0.1	0.25	Median	0.75	0.9	Fixed
Earnings							
Under jun. high school	-.2631 -11.04	-.33455 -8.62	-.2532 -7.89	-.2295 -9.49	-.24764 -9.92	-.26234 -7.21	-.03395 -0.19
Junior high school	-.13708 -8.03	-.14216 -4.82	-.11518 -5.01	-.14753 -8.70	-.15932 -8.98	-.15932 -6.37	.059085 0.53
Junior college	.148031 9.64	.200922 7.29	.155521 7.61	.115015 7.62	.101667 6.53	.080631 3.64	.182208 3.09
University+	.343783 24.05	.398457 15.44	.322806 16.99	.300276 21.41	.312496 21.55	.29091 13.93	.189115 3.41
Training	.064931 4.53	.099107 4.06	.085776 4.49	.061891 4.34	.044469 3.06	.03651 1.74	.006228 0.56
Observations	7359 (Groups 2709)						
(Psd.) Rsq	0.5417	0.3305	0.3367	0.3540	0.3778	0.3806	0.3231

Note *Base variable for education level: Senior high school. t-statistics are presented in the 2nd. line of cells in the table.*

Table 2.10: Estimation Results using Variables - Educational Level and Job Training in Germany

Germany	OLS	0.1	0.25	Median	0.75	0.9	Fixed
Earnings							
Under Real/ Hauptschule	-.3801 -12.28	-.70773 -20.09	-.39373 -12.14	-.26174 -10.10	-.14059 -3.77	-.1822 -4.11	-.49324 -5.72
Real/ Hauptschule	-.19373 -16.13	-.15795 -8.37	-.16376 -12.95	-.15248 -15.95	-.13335 -9.96	-.13436 -8.52	-.40937 -13.64
Technik/ Meister-schul	.146603 9.36	.131919 5.48	.120961 7.39	.14112 11.64	.145554 8.91	.161465 8.24	.138322 1.77
Fachhochschule+	.520474 35.02	.501617 20.70	.490792 30.88	.509401 44.31	.486228 31.88	.476256 25.76	1.396115 16.45
Training	.165382 4.51	.078855 1.40	.165109 4.21	.172733 6.04	.094636 2.60	.057429 1.32	.037328 1.45
Observations	13011 (Groups 4129)						
(Psd.) Rsq	0.5825	0.4819	0.4024	0.3434	0.3200	0.3104	(g:4129) 0.1368

Note *Base variable for education level: Berufsabschluss and/or (Fach-) Hochschulreife. t-statistics are presented in the 2nd. line of cells in the table.*

Table 2.11: Estimation Results using Variables - Educational Years and Job Training in Korea

Korea	OLS	0.1	0.25	Median	0.75	0.90	Fixed
Earnings							
0 years	-.4646 -8.85	-.43452 -3.90	-.4865 -8.26	-.46591 -8.05	-.56595 -11.08	-.49611 -6.87	(dropped)
6 years	-.22481 -8.72	-.23806 -4.40	-.18624 -6.27	-.19061 -6.53	-.22531 -7.84	-.23604 -6.34	-.03406 -0.19
9 years	-.13768 -8.11	-.14352 -4.12	-.11827 -6.15	-.14666 -7.80	-.15652 -8.34	-.16297 -6.86	.059117 0.53
14 years	.145876 9.56	.203711 6.32	.15671 9.11	.112176 6.71	.09584 5.74	.079104 3.76	.18231 3.09
16 years	.311178 21.51	.385218 12.34	.308679 18.83	.26033 16.47	.27316 17.22	.272423 13.54	.189326 3.41
18+ years	.595387 18.95	.704542 10.47	.623732 17.52	.539342 15.70	.552695 16.01	.563908 12.87	.239661 2.54
Training	.064447 4.53	.082036 2.84	.081018 5.04	.067302 4.24	.046106 2.94	.032856 1.66	.006148 0.55
Obs.	7359 (Groups 2709)						
(Psd.) sq	0.5472	0.3338	0.3422	0.3604	0.3832	0.3338	0.3231

Note *Base variable for educational years: 12 years education. t-statistics are presented in the 2nd. line of cells in the table.*

Figure 2.3: Effects of Educational Years and Job Training in Korea

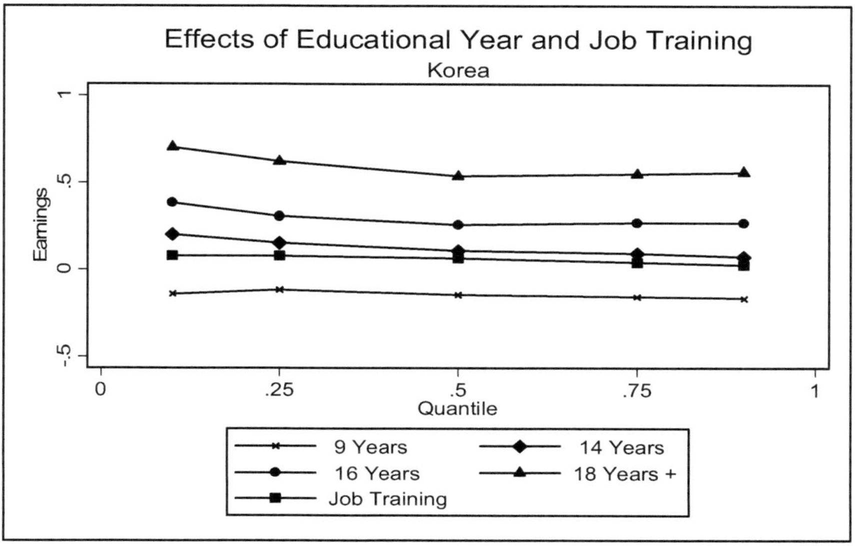

Figure 2.4: Effects of Educational Years and Job Training in Germany

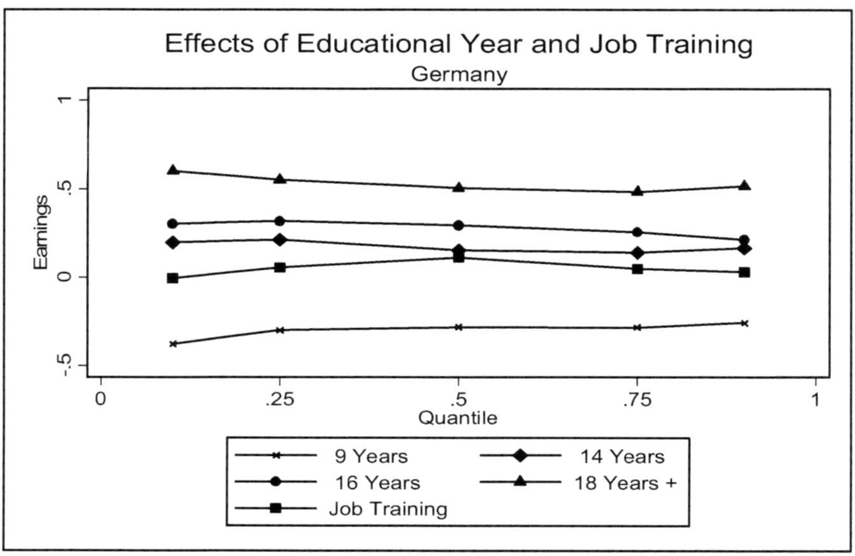

Table 2.12: Estimation Results using Variables - Educational Years and Job Training in Germany

Germany	OLS	0.1	0.25	Median	0.75	0.90	Fixed
				Earnings			
7 years	-.45541 -12.56	-.78308 -21.34	-.45483 -13.32	-.3285 -10.39	-.26793 -7.24	-.2676 -5.38	-. 9345 -6.79
8.5 years	-.17027 -2.45	-.08998 -0.92	-.17632 -2.73	-.20947 -3.92	-.17308 -2.82	-.10743 -1.36	-. 93149 -5.01
9 years	-.32381 -17.29	-.37879 -14.44	-.30111 -16.93	-.28213 -18.45	-.28412 -16.13	-.25759 -10.95	-.32127 -2.12
10 years	-.28548 -12.68	-.35288 -11.13	-.34441 -16.99	-.34992 -20.46	-.22168 -9.91	-.17764 -5.82	-.2361 -2.36
10.5 years	-.12579 -7.70	-.07353 -2.93	-.10112 -6.36	-.12736 -9.83	-.14491 -9.90	-.14499 -7.45	. 390076 1.84
11 years	-.08907 -4.12	-.02382 -0.71	-.0454 -2.14	-.08586 -5.00	-.09956 -5.13	-.07432 -2.84	. 546063 2.29
11.5 years	.001707 0.09	.077453 2.71	.010253 0.58	-.01634 -1.14	-.01339 -0.82	-.00352 -0.16	.270907 2.14
13 years	-.20314 -8.37	-.36934 -11.56	-.37387 -18.39	-.18348 -9.40	.058984 2.28	.173279 4.76	.953721 4.50
13.5 years	.02636 0.82	.091144 1.85	.082163 2.58	.044798 1.73	.028692 0.99	.068307 1.76	.148768 1.42
14 years	.152465 4.21	.199159 3.59	.213625 6.03	.156777 5.45	.143282 4.42	.167471 3.87	(dropped)
14.5 years	.048352 1.80	.154347 3.63	.100055 3.76	.086982 4.00	.088613 3.59	.062302 1.92	1.269924 5.59
15 years	.27483 10.37	.314978 7.77	.301258 11.71	.287014 13.58	.300337 12.39	.273002 8.66	.675298 4.52
16 years	.270203 9.06	.305133 6.77	.320063 10.96	.295814 12.39	.258734 9.61	.216388 6.19	2.965298 8.52
17 years	.178832 2.35	.050905 0.50	.232346 3.21	.303257 4.80	.277797 4.10	.265148 2.95	(dropped)
18+ years	.549202 21.43	.603304 15.54	.553175 22.52	.506652 25.29	.486639 21.35	.518342 16.72	7.116118 13.30
Training	.101212 2.72	-.00604 -0.11	.05585 1.60	.113542 3.81	.049917 1.54	.033897 0.78	.001325 0.05
Obs.				12934 (Groups 4089)			
(Psd.) Rsq	0.5918	0.4942	0.4162	0.3534	0.3301	0.3205	0.1541

Note *Base variable for educational years: 12 years education. t-statistics are Presented in the 2nd. line of cells in the table.*

Table 2.13: Job Training Effects on Earnings including Interaction Terms (Educational Levels and Job Training) in Korea

Korea	OLS	0.1	0.25	Median	0.75	0.9	Fixed
Earnings							
Edu1: Under J. high sch.	-.27884 -11.26	-.32609 -6.83	-.26976 -8.77	-.25241 10.25	-.28318 10.76	-.27838 -8.12	-.05311 -0.29
Edu2: Junior high school	-.13885 -7.88	-.14256 -4.13	-.11524 -5.25	-.14361 -8.35	-.15694 -8.41	-.15864 -6.53	.026599 0.24
Edu4: Junior college	.16447 9.90	.236845 6.90	.174185 8.55	.128782 8.05	.112686 6.47	.094895 4.35	.17422 2.95
Edu5: University+	.355283 22.98	.42079 13.08	.341778 17.83	.310329 20.84	.313352 19.48	.313259 15.27	.187833 3.37
Completed training	.099383 4.29	.155623 3.35	.120797 4.16	.10241 4.51	.067233 2.80	.083734 2.68	-.00972 -0.53
Edu1*comp. training	.257677 2.57	.109936 1.10	.00761 0.12	.006375 0.13	-.04031 -0.77	-.05201 -0.75	-.06018 -1.59
Edu2*comp. training	.078497 1.06	.252332 1.32	.246222 2.08	.177488 1.90	.386245 3.57	.131986 1.07	.121974 1.40
Edu4*comp. training	-.09595 -2.57	.115555 0.82	.042274 0.47	.019813 0.28	.032424 0.42	.038647 0.41	.126851 1.83
Edu5*comp. training	-.05293 -1.85	-.13628 -1.89	-.06171 -1.33	-.0762 -2.06	-.06628 -1.66	-.0725 -1.44	.038823 1.20
(Psd.) Rsq	0.5429	0.3322	0.3374	0.3547	0.3788	0.3822	0.3239

Note *Base variable for education level: Senior high school. Base variable for interaction terms: Completed job training* Senior high school. t-statistics are presented in the 2nd. line of cells in the table.*

Table 2.14: Job Training Effects on Earnings including Interaction Terms (Educational Levels and Job Training) in Germany

Germany	OLS	0.1	0.25	Median	0.75	0.9	Fixed
Earnings							
Edu1: Under Real/Haupt	-.37935 -12.25	-.703867 -19.74	-.39364 -13.17	-.26418 -9.55	-.13965 -4.11	-.18126 -4.15	-.49341 -5.71
Edu2: Real/ Hauptschul	-.19258 -15.98	-.159242 -8.40	-.1589 -13.69	-.15272 -14.96	-.13288 -10.87	-.13615 -8.50	-.40956 -13.52
Edu4:Tecnik/ Meisterschul	.148862 9.46	.1345278 5.50	.121808 8.04	.141637 10.92	.148818 9.96	.16324 8.20	.137095 1.75
Edu5: Fach-hochschul+	.522524 35.11	.5062932 20.66	.494106 33.65	.511226 41.66	.488704 35.20	.477322 25.56	1.39582 16.45
Completed training	.198228 5.01	.1415296 2.28	.242657 6.09	.227623 6.88	.102882 2.90	.046813 1.01	.036492 1.31
Edu1*comp. training	dropped	dropped	dropped	dropped	dropped	dropped	dropped
Edu2*comp. training	.096553 0.39	.0549977 0.41	-.09654 -0.45	.219424 1.15	.179844 0.76	.023776 0.26	.050569 0.34
Edu4*comp. training	-.19456 -1.47	-.372237 -2.06	-.21807 -1.90	-.20928 -2.01	-.10927 -0.92	-.1614 -0.98	.021154 0.21
Edu5*comp. training	-.37935 -2.62	-.669307 -10.18	-.40862 -3.56	-.21194 -1.70	-.25061 -2.02	-.34238 -4.87	-.05046 -0.41
(Psd.) Rsq	0.5828	0.4824	0.4026	0.3437	0.3203	0.3107	0.1369

Note *Base variable for education level: Berufsabschluss and/or (Fach-) Hochschulreife. Base variable interaction terms: Completed job training*Berufsabschluss and/or (Fach-)Hochschulreife. t-statistics are presented in the 2nd. line of cells in the table.*

Table 2.15: Estimation Results using Variables: Job Training in Earlier Years

	Germany			Korea		
	OLS	**Median**	**Fixed**	**OLS**	**Median**	**Fixed**
			Earnings			
Edu1	-.30594 -4.30	-.36596 -6.43	(dropped)	-.29163 -7.60	-.21973 -4.41	(dropped)
Edu2	-.09665 -4.64	-.05371 -2.98	-.71781 -4.38	-.09927 -3.31	-.09157 -2.44	(dropped)
Edu4	.156893 6.70	.157349 7.95	.231754 0.85	.144574 4.86	.147361 3.96	-.14788 -1.19
Edu5	.530716 22.93	.547663 27.87	.823846 6.29	.334746 12.36	.307347 9.19	-.05588 -0.46
Training Last year	.007701 0.10	.006342 0.10	.02673 0.42	.063313 2.43	.079998 2.45	-.02414 -0.88
Train -2 year	.019255 0.33	.010003 0.21	.008535 0.15	.013355 0.50	.01509 0.45	.007 0.23
Train - 3 year	.145141 2.60	.058145 1.28	.009238 0.18	.036175 1.28	.016453 0.47	.001813 0.05
Train - 4 year	.13247 2.39	.040666 0.91	-.03978 -0.82	.033232 1.12	.054994 1.47	.037101 1.08
Train - 5 year	.169354 3.03	.128068 2.76	-.01798 -0.45	.077932 2.59	.064968 1.75	.023256 0.80
Train -5 year+	-	-	-	.057191 2.00	. 057512 1.62	(dropped)
Train - 6 year	.027818 0.58	.054325 1.32	-.06369 -1.88	-	-	-
Train - 7 year	. 074594 1.71	. 049597 1.38	-.0395 -1.37	-	-	-
Obs.	4018	4018	4018/2082	1711	1711	1711/1041
(Psd.) Rsq	53.55	0.3051	0.1018	0.5715	0.3700	0.3087

Note *Base variable for education level: Berufsabschluss and/or (Fach-) Hochschulreife for Germany and Senior high school for Korea. t-statistics are presented in the 2nd. line of cells in the table.*

Table 2.16: Estimation Results - Age and Tenure in Korea

Korea	OLS	0.1	0.25	Median	0.75	0.9	Fixed
				Earnings			
Age	.084944 13.70	.098241 7.39	.102846 12.86	.075045 12.40	.062575 9.71	.064827 7.78	.062403 0.46
Age squared	-.000961 -12.66	-.001197 -7.49	-.001232 12.81	-.000859 -11.54	-.000662 -8.26	-.000631 -6.08	-.001219 -6.82
Tenure	.034241 14.69	.049698 10.38	.033558 11.51	.031835 13.96	.025718 10.27	.023794 6.82	-.011358 -3.02
Tenure squared	-.000537 -5.72	-.001097 -6.08	-.000467 -4.03	-.000398 -4.33	-.000238 -2.32	-.000263 -1.80	.000733 3.61

Note *t-statistics are presented in the 2nd. line of cells in the table.*

Table 2.17: Estimation Results - Age and Tenure in Germany

Germany	OLS	0.1	0.25	Median	0.75	0.9	Fixed
				Earnings			
Age	.135165 31.58	.185469 27.23	.151455 36.04	.128704 36.43	.088314 22.64	.072898 14.22	.120327 11.22
Age squared	-.001418 -27.58	-.001940 -24.47	-.001597 -32.04	-.001354 -31.80	-.000895 -18.59	-.000685 -10.74	-.001157 -9.07
Tenure	.015821 9.76	.022065 8.98	.016024 10.48	.009347 6.98	.010323 6.55	.010726 4.89	-.008068 -2.87
Tenure squared	-.000293 -5.94	-.000405 -5.71	-.000307 -6.83	-.000142 -3.48	-.000192 -3.93	-.000275 -4.06	.000283 2.66

Note *t-statistics are presented in the 2nd. line of cells in the table.*

Figure 2.5: Age-Earnings Profiles in Korea

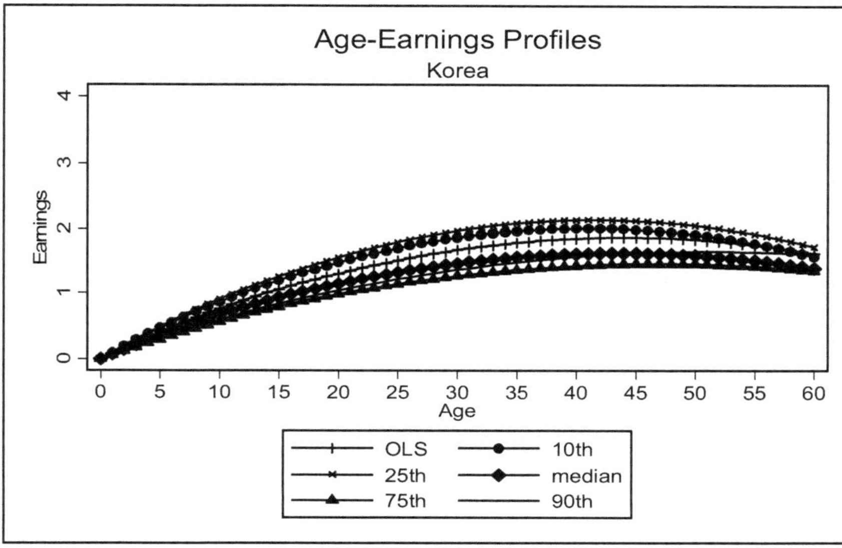

Figure 2.6: Age-Earnings Profiles in Germany

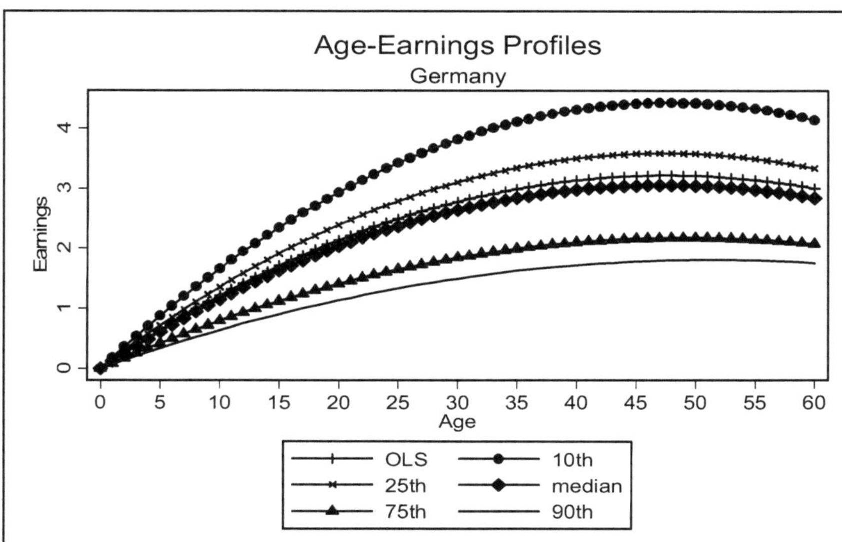

Figure 2.7: Tenure-Earnings Profiles in Korea

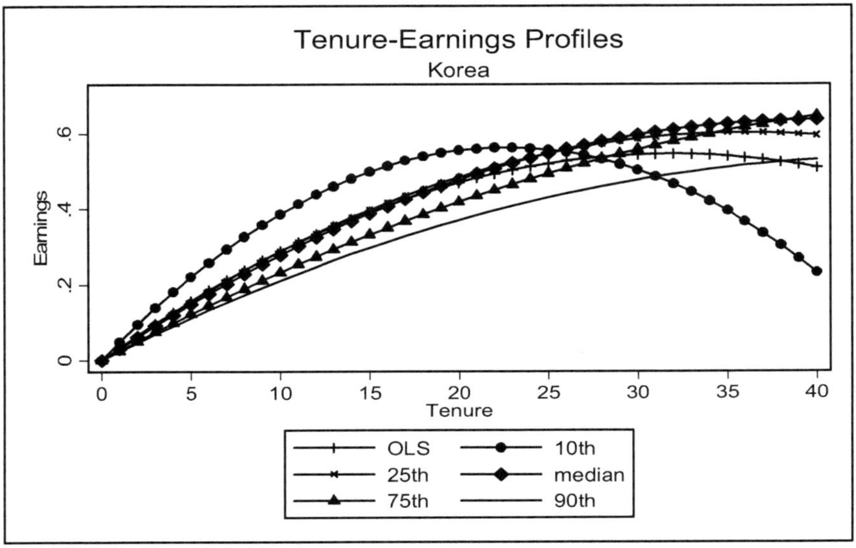

Figure 2.8: Tenure-Earnings Profiles in Germany

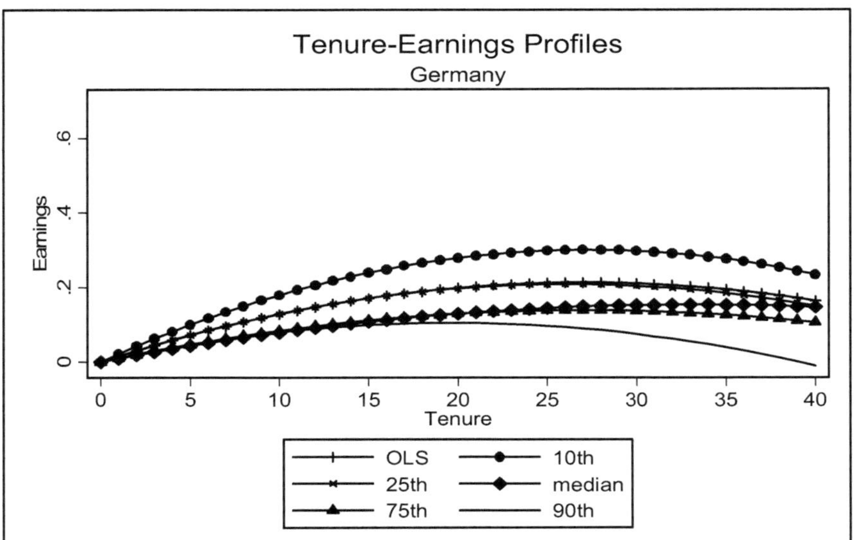

Table 2.18: Estimation Results - Family Characteristics in Korea

Korea	OLS	0.1	0.25	Median	0.75	0.9	Fixed
Earnings							
Married	.080002 4.03	.167532 4.05	.103345 4.17	.062893 3.26	.028581 1.40	.01554 0.59	-.05482 -1.70
Widowed	-.06021 -0.66	-.02851 -0.20	-.12484 -1.23	-.03484 -0.40	.002714 0.03	-.0168 -0.18	-.05269 -0.39
Divorced	-.04881 -1.35	-.08011 -1.13	-.06798 -1.55	.042318 1.14	.014673 0.37	.008344 0.17	-.0611 -0.87
Separated	-.01079 -0.15	.061915 0.43	-.06226 -0.72	-.03274 -0.48	-.06711 -0.91	. 025888 0.26	-.02453 -0.28
Household head	.09825 5.88	.064782 2.09	.069466 3.50	.093812 5.74	.097044 5.41	.115864 4.90	.090347 2.85
Children	.021439 1.60	-.01713 -0.62	.015499 0.92	.032329 2.45	.067175 4.80	.071237 4.14	.002513 0.15

Note *Base variable for marital status: Single. t-statistics are presented in the 2nd. line of cells in the table.*

Table 2.19: Estimation Results - Family Characteristics in Germany

Germany	OLS	0.1	0.25	Median	0.75	0.9	Fixed
Earnings							
Married	.123301 8.53	.185268 8.37	.117851 8.60	.072356 6.19	.060042 4.53	.048208 2.74	.000982 0.04
Widowed	-.00908 -0.10	-.10741 -0.90	.055571 0.64	.015727 0.21	-.04385 -0.53	-.06119 -0.63	-.12016 -1.17
Divorced	.075625 3.42	.136079 3.80	.063495 2.98	.027351 1.53	.026865 1.32	.039889 1.45	-.02685 -0.74
Separated	.032322 0.91	.158781 2.71	.038341 1.11	-.00789 -0.28	-.03846 -1.22	-.05621 -1.34	-.00695 -0.19
Household head	.032456 3.30	.032514 2.13	.056203 5.89	.035279 4.32	.021733 2.31	.007936 0.64	.034443 1.53
Children	.002253 0.22	.030119 1.75	.003658 0.35	.004135 0.48	.019332 1.99	.027931 2.16	.006309 0.45

Note *Base variable for marital status: Single. t-statistics are presented in the 2nd. line of cells in the table.*

Table 2.20: Estimation Results - Firm Characteristics in Korea

Korea	OLS	0.1	0.25	Median	0.75	0.9	Fixed
			Earnings				
Public	.02003 0.81	.02076 0.43	.02327 0.76	.01863 0.78	.01861 0.73	-.00096 -0.03	-.02548 -0.86
Overtime	.09255 8.56	.11756 5.45	.10256 7.70	.08906 8.43	.07021 6.10	.05226 3.57	.00735 0.84
Part time	-.3510 -12.11	-.5824 -13.05	-.4181 -12.40	-.3034 -10.40	-.2003 -5.78	-.1548 -3.60	-.1779 -4.36
Firm size 5-9	.02401 1.19	.05373 1.34	.05475 2.17	.01927 0.98	.02791 1.29	.0277 1.02	.01638 0.85
Firm size 10-99	.11957 6.72	.19003 5.29	.1598 7.11	.09541 5.54	.0834 4.50	.10515 4.42	.10515 2.64
Firm size 100-999	.18452 9.07	.22841 5.56	.20492 8.01	.14907 7.60	.1668 7.81	.20028 7.39	.07910 3.49
Firm size 1000+	.36986 16.00	.43497 9.01	.41083 13.88	.32923 14.77	.35198 14.61	.36497 11.91	.11977 4.65

Note *Base variable for firm size: Firm with 1 to 4 employees. t-statistics are presented in the 2nd. line of cells in the table.*

Table 2.21: Estimation Results - Firm Characteristics in Germany

Germany	OLS	0.1	0.25	Median	0.75	0.9	Fixed
Earnings							
Public	-.05012 -3.25	.023888 0.86	-.00063 -0.04	-.07008 -5.55	-.11992 -8.43	-.17761 -9.28	.03407 1.45
Overtime	.145994 15.65	.179491 12.33	.11755 13.23	.107921 14.20	.098454 11.33	.099525 8.62	.035073 4.80
Part time	-.32091 -28.13	-.51438 -33.57	-.38902 -37.32	-.24465 -24.62	-.18694 -15.72	-.18527 -11.74	-.02244 -1.93
Firm size 5-19	.042141 1.86	.150886 3.97	.118118 5.13	.050479 2.67	-.00579 -0.28	.030373 1.06	.003355 0.13
Firm size 20-99	.164704 6.96	.272687 6.85	.24699 10.25	.192679 9.71	.088625 4.12	.068729 2.40	.061863 2.06
Firm size 100-1999	.209967 8.83	.342004 8.50	.296387 12.28	.233572 11.73	.1413 6.48	.133391 4.51	.069026 2.27
Firm size 2000+	.298349 11.80	.432705 10.28	.370557 14.53	.301021 14.34	.211468 9.19	.199558 6.44	.081137 2.53

Note *Base variable for firm size: Firm with 1 to 4 employees. t-statistics are presented in the 2nd. line of cells in the table.*

Table 2.22: Estimation Results – Family Residence in Korea

Korea	OLS	0.1	0.25	Median	0.75	0.9	Fixed
Earnings							
Busan	-.10453 -5.99	-.0801 -2.23	-.07731 -3.50	-.08369 -4.84	-.07364 -3.91	-.11182 -4.83	-.08654 -0.91
Daegu	-.16213 -8.53	-.14491 -3.79	-.1417 -5.94	-.15107 -8.05	-.17011 -8.45	-.18903 -7.46	.08258 0.76
Deajeon	-.07531 -2.78	.016649 0.30	-.04369 -1.31	-.09892 -3.80	-.10304 -3.60	-.14624 -4.13	.098776 1.10
Incheon	-.09082 -4.63	-.103 -2.66	-.1019 -4.24	-.06184 -3.17	-.07959 -3.77	-.09379 -3.53	.018599 0.27
Gwangju	-.16189 -6.10	-.19239 -3.72	-.18343 -5.65	-.12549 -4.74	-.14867 -5.25	-.17066 -4.79	-.12014 -1.33
Ulsan	-.02577 -1.07	-.00375 -0.08	-.00772 -0.25	-.01695 -0.72	-.07456 -3.05	-.12343 -4.06	.040251 0.33
Gyunggi	-.03559 -2.53	-.02836 -1.01	-.01766 -1.01	-.01849 -1.33	-.00654 -0.43	-.03157 -1.63	.053722 1.37
Gangwon	-.08041 -1.88	-.01529 -0.18	-.12282 -2.45	-.09394 -2.27	-.10594 -2.39	-.07989 -1.41	-.09357 -0.76
Chungbuk	-.10213 -3.18	-.00746 -0.11	-.02816 -0.67	-.10939 -3.52	-.13376 -4.03	-.19193 -4.82	-.07049 -0.56
Chungnam	-.10657 -3.93	-.08106 -1.50	-.04996 -1.46	-.08128 -3.03	-.10808 -3.75	-.18704 -5.38	.044383 0.58
Jeonbuk	-.11473 -3.98	-.0501 -0.88	-.11236 -3.16	-.13853 -4.99	-.13676 -4.51	-.13117 -3.37	. 318487 3.41
Jeonnam	-.03847 -1.20	-.02257 -0.36	-.05878 -1.50	-.03405 -1.08	. 009034 0.26	-.04515 -1.07	-.06568 -0.56
Gyungbuk	-.19307 -8.22	-.24818 -5.45	-.20199 -7.01	-.15603 -6.67	-.18127 -7.20	-.13415 -4.10	-.18594 -2.07
Gyungnam	-.08482 -4.62	-.02884 -0.75	-.05544 -2.40	-.09595 -5.37	-.11829 -6.22	-.10669 -4.41	-.0327 -0.37
Jeju	.693438 1.34	2.118955 19.79	1.157968 21.60	.589985 16.27	.278378 7.89	.013408 0.35	.404385 0.97

Note *Base variable for residence: Capital city Seoul. t-statistics are presented in the 2nd. line of cells in the table.*

Table 2.23: Estimation Results – Family Residence in Germany

Germany	OLS	0.1	0.25	Median	0.75	0.9	Fixed
Earnings							
Schleswig-Hols	.088965 2.14	.048422 0.73	.088021 2.16	.0923 2.69	.03628 0.95	-.15749 -3.46	.225761 1.07
Hamburg	.144997 2.92	.062988 0.83	.10987 2.30	.11491 2.84	.054988 1.22	-.12728 -2.34	.215865 0.88
Lower Saxony	.107489 2.93	.056853 0.99	.073173 2.07	.093095 3.10	.06671 1.96	-.07868 -1.90	.007795 0.06
Bremen	.127184 2.04	.102038 1.07	.136071 2.23	.095576 1.89	.113521 1.96	-.03901 -0.54	-.0867 -0.54
N-Rhein-Westfa	.146174 4.08	.10229 1.81	.089238 2.62	.125457 4.30	.095351 2.88	-.04835 -1.21	.153404 1.14
Hessen	.19495 5.02	.114067 1.90	.12848 3.49	.137036 4.39	.168814 4.64	.009137 0.21	.149978 1.14
R-Pfalz,Saarl.	.116832 3.09	.116315 1.92	.083524 2.30	.102011 3.29	.072206 2.05	-.09665 -2.28	.031727 0.24
Baden-Wuerttemb.	.226734 5.98	.165713 2.79	.165658 4.59	.186174 6.07	.160283 4.60	-.0113 -0.27	.217186 1.70
Bavaria	.170797 4.66	.128271 2.22	.092364 2.67	.137095 4.61	.107782 3.19	-.0592 -1.46	.365655 2.21
Berlin (East)	.061834 0.85	-.02324 -0.22	.007311 0.11	-.08292 -1.49	.104403 1.49	.007616 0.09	-.02805 -0.20
Mecklenburg-V.	.038628 0.61	-.26269 -3.02	-.19597 -3.55	.056886 1.08	-.06287 -1.10	-.10607 -1.50	.321711 0.77
Brandenburg	-.13161 -2.94	-.35927 -5.74	-.22335 -5.38	-.06618 -1.73	-.11701 -2.74	-.1708 -3.10	-.18391 -1.89
Saxony-Anhalt	-.03437 -0.66	-.2299 -3.07	-.03393 -0.67	.051124 1.14	.013529 0.27	-.17318 -2.83	-.042 -0.29
Thueringen	-.09015 -1.82	-.22351 -3.04	-.1814 -3.95	-.09919 -2.44	-.11677 -2.53	-.20462 -3.59	.197757 0.99
Saxony	-.09349- 2.18	-.12765 -1.89	-.17118 -4.27	-.11519 -3.30	-.1768 -4.60	-.25912 -5.40	-.01038 -0.06

Note *Base variable for residence: Berlin (West). t-statistics are presented in the 2nd. line of cells in the table.*

Figure 2.9: Earnings Effects of Marriage

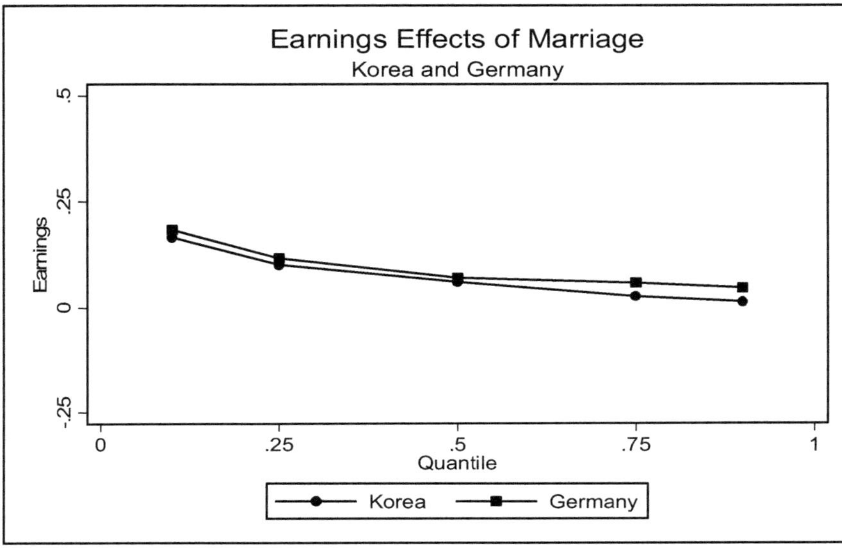

Figure 2.10: Earnings Effects of Household Head

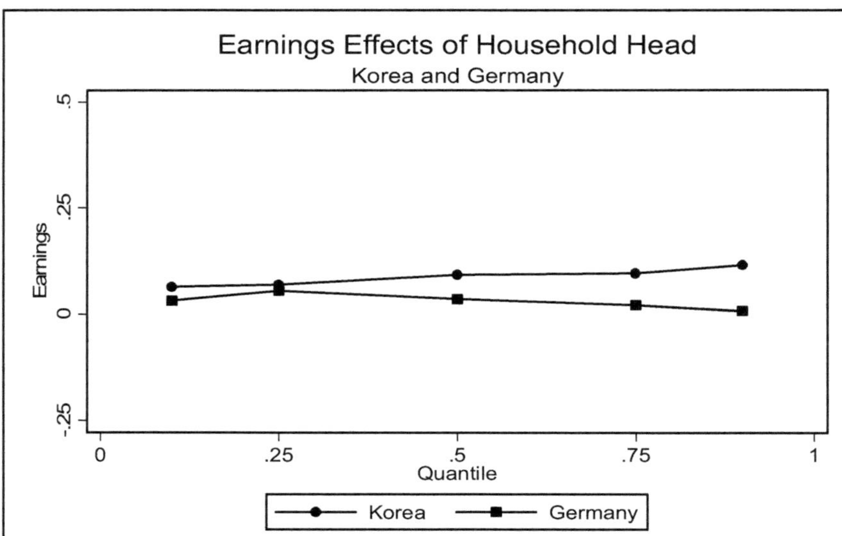

Figure 2.11: Earnings Effects of Having Children

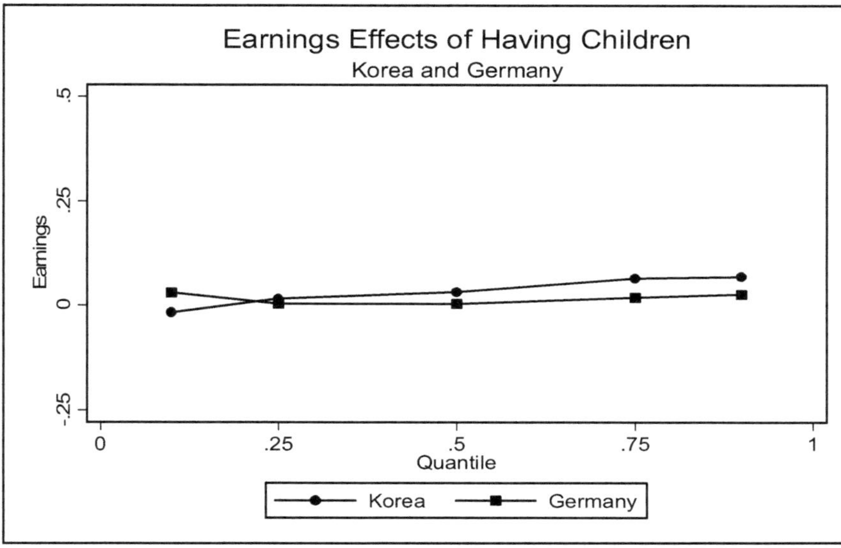

Figure 2.12: Earnings Effects of Working in Public Sector

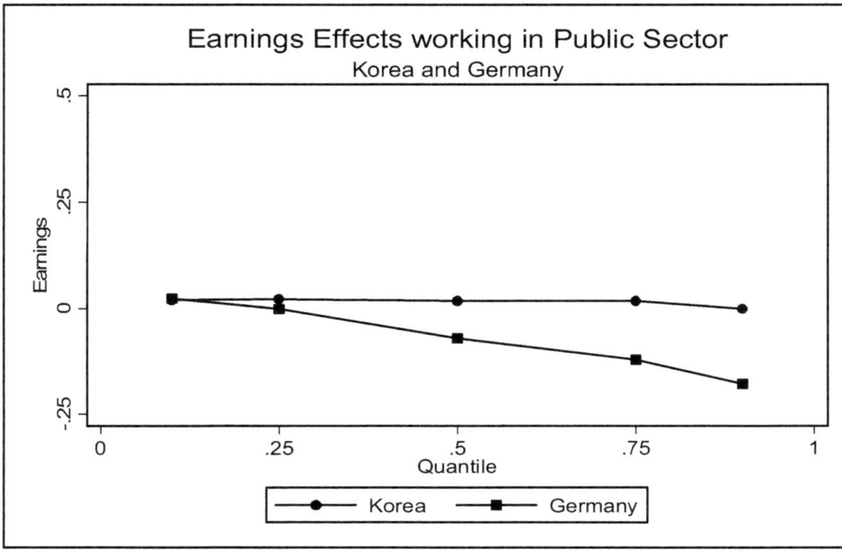

Figure 2.13: Earnings Effects of Working Overtime

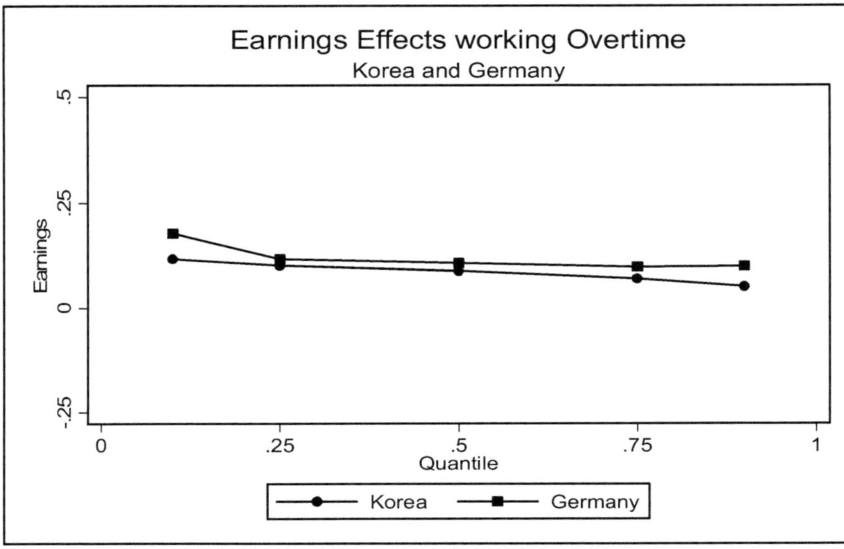

Figure 2.14: Earnings Effects of Working Part Time

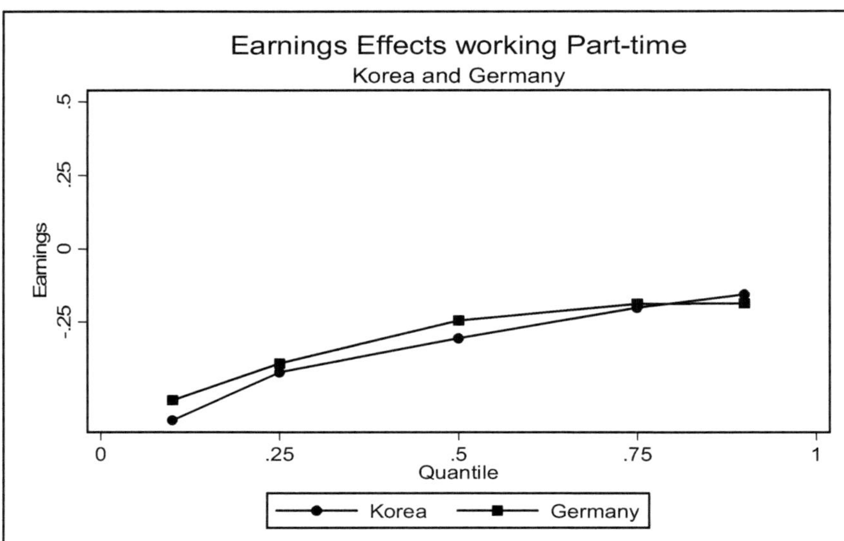

Figure 2.15: Earnings Effects of Firm Size by Employee in Korea

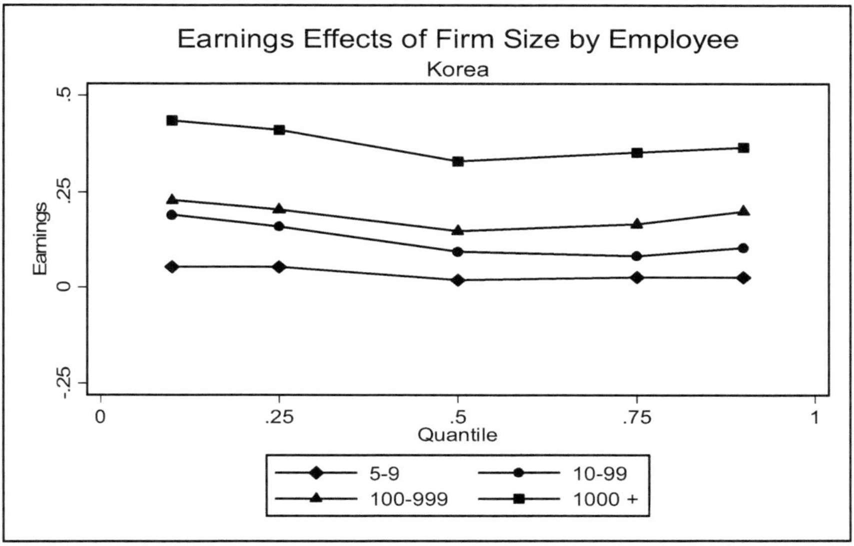

Figure 2.16: Earnings Effects of Firm Size by Employee in Germany

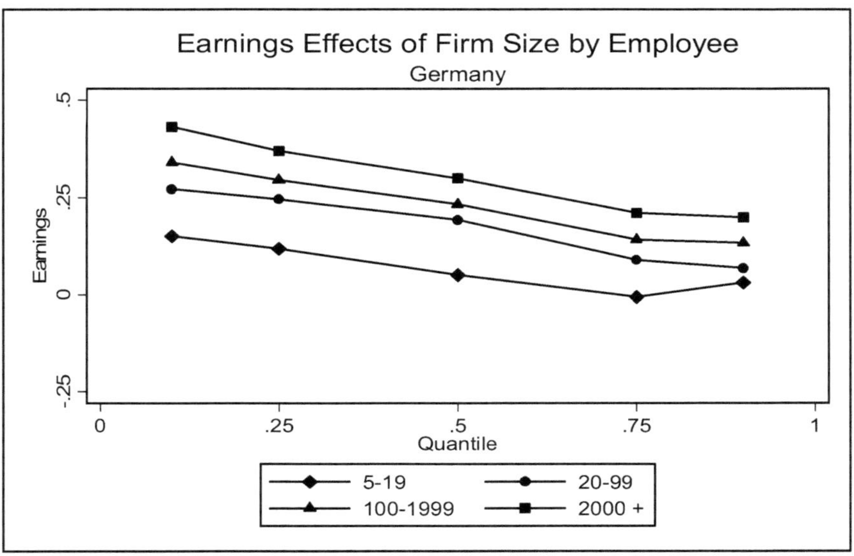

3. Women's Career and Children

– Human Capital and Earnings of Women in Korea and Germany. A Comparative Study

Abstract

According to the theory of Mincer and Polachek (1974), this study tests the career discontinuity of women, measured by their earnings, due to childbirth and childcare. Using the data taken from the 'Korean Labor and Income Panel Study (KLIPS)' and 'German Socio-Economic Panel (GSOEP),' this study estimates the effects of human capital and having children on women's earnings by means of the regression discontinuity design and quantile regression methods. This study found earnings discontinuities in women aged between 30 and 40 in Germany and women aged between 40 and 55 in Korea. German women suffer from a more than 65% earnings drop due to children at their average tenure of 7 years. Unexpectedly, Korean women benefit from a 37% earnings gain attributable to children at their average tenure of 4 years.

Keywords: Women, Education, Family, Children, Human capital, Earnings

3.1. Introduction

The increasing importance of education in the labor market has proved a significant motivation for (ambitious) women to acquire a higher level of education. On the other hand, caring for children and family is recognized as a hindrance for a women's career progression in the labor market. In human capital theory, childcare is formulated as a determinant, which reduces women's productivity and earnings. This brings about a discontinuity in women's working experience and depreciates their skills (Mincer and Polachek (1974)). Numerous studies[20] demonstrate that women rearing children have, on average, lower earnings than women without children. This empirical fact is described as the earnings penalty of women. Reasons behind this penalty include not only women's career inter-

20 Mincer and Polachek (1978); Mincer and Ofek (1982); Licht and Steiner (1991); Polachek and Kim (1994); Light and Ureta (1995), and many others.

mittence due to child birth and child rearing, but also the occupational segregation of women in lower paying jobs and the difficulties mothers experience when working on a full time basis. However, low effort and the weak attachment to the labor market of working mothers are also pointed out as possible reasons for women's low earnings (Becker (1985)).

Although certain family friendly policies sometimes encourage working women to have children, rearing children accompanied with success on the job still seems to be a big challenge for women in many countries. In Germany, the total fertility rate was measured as 1.34 children per woman in 2005. The fertility rate was 1.08 children in Korea in the same year. Korea thus recorded the lowest birth rate of any country in the OECD (see OECD (2007)). The low birth rate of (qualified) women in Germany and Korea might reflect the difficulties in balancing work and family life for women. How attractive or how difficult is having a baby for working women? This study, according to the theories of Mincer and Polachek (1974, 1978), hypothesizes a discontinuity in women's careers - measured by their earnings - due to children and predicts the earnings penalty of working mothers in Korea and in Germany. Testing their theories, the study attempts to determine the effect having children has on women's earnings in Korea and to compare this effect with the situation in Germany, which is taken from the same empirical framework.

Reviewing the literature, in Germany, Galler (1991), Licht and Steiner (1992) studied the effects of an employment break experienced by German women on their income profile. Galler found a strong income reduction for women due to an employment break. 10 years break reduces potential life income by 1/3. Licht and Steiner also found a persistent income penalty due to any employment breaks for women. But according to their study, this penalty is partly compensated for by restoring their depreciations of skills after reintegration in the labor market. Beblo and Wolf (2000, 2002) found that the depreciation rates of women's human capital due to an employment break are substantially affected by the type of employment under consideration, such as industrial sector, job position and working part time or full time in Germany. Their study demonstrated that not only the duration of the break, but also its timing plays an important role in any earnings loss suffered by women. Beblo and Wolf (2002) further differentiated the wage effects of employment breaks between registered unemployment, formal parental leave, training and other reasons for both women and men. According to the study, career interruptions reduce the wage rates of both men and women, but this reduction of wages between women and men is very different in sign and size. While unemployment and out-of-the-labor-force periods mainly lead to wage reductions for men, wage cuts for women can mainly be ascribed to parental leave and additional home time, even if this parental leave and the subsequent home time took place several years ago. Unemployment leads to a lower wage cut for women than staying out of the labor force for a while. Kunze (2002) distinguished four types of career interruptions – unemployment, parental leave for female workers, national service for male workers and other non-work spells. She analyzed their short and long term effects on wages. Kunze found that timing effects seem important, and net depreciation differs according to the type of interruption.

Concerning maternity leave, Ondrich, Spiess and Yang (2001) found that maternity leave has a negative effect on wage growth in Germany: Each month of maternity leave reduces wage growth by 1.5 percent as measured over five years (1984-1989 and 1989-1994). Six months out of the labor force after the end of the parental leave period lowers wage growth by an additional 15 percent as measured over five years between 1989 and 1994. Ejrnaes and Kunze (2004) estimated the wage effects for female workers around the time of the birth of their first child. Daily real wages reveal a dip shortly before giving birth and a drop of 10 to 20 percent after finishing maternity leave and returning to the labor market. Beblo, Bender and Wolf (2006) analyzed wage effects of employment breaks on women during motherhood by means of a within-firm matching approach. They found that the first birth reduced women's wages by 16 to 19 percent, and the wage loss increased during the employment break after childbirth. Schönberg and Ludsteck (2007) analyzed the impact of expansions in leave coverage on mothers' labor market outcomes after childbirth. They found that each expansion induced women to delay their return to work. The expansion in leave coverage from 2 to 6 months reduced women's wages even 8 years after childbirth, which could mostly be explained by the delay in returning to work.

In Korea, empirical studies since the 1980's concerning women in the labor market have primarily focused on gender earnings inequality using decomposition methods.[21] In the 2000's, researchers in Korea concentrated on the employment and wage effects of women's working conditions such as working on a part/full time basis, working under temporary contract or shortening working hours (Kim (2001), Moon and Kang (2003) and Kang (2004)). At the end of the 2000's, the low birth rate and working conditions became the focus of

21 Park (1982), Yun (1990), Huh and Hong (2003), Shin (2003), Kim (2007), Kang (2008) and many others.

research on women. Park (2008) analyzed, through a questionnaire for employees in Seoul, the psychological burden of childcare. This study found that the psychological burden of childcare was higher for women than for men and for singles than for married couples. Childless people imagine the burden would be higher than people with children. People with children think about childbirth more positively than people with no children. Married people consider parental leave to be the most important family friendly employment policy, while singles think childcare services at work are more important. Concerning wage levels and birth rates, Kim, J.H. (2009) determined that a 10% wage increase for a male leads to a 1 to 3.2% positive effect on the probability of having a second child birth, while the same increase for a female leads to a -1.6 to -2.7% negative effect. This wage increase for women explains the 17% decrease in the probability of having a second child using the period model. Concerning working conditions and birth rate, Kim, S.H. (2009) found that women working on a part time basis with a temporary contract or as casual workers have a lower birthrate than women working on a full time basis or as regular employees. Women working in the manufacturing industry have a lower birthrate than in other industries. Further, the childcare offered has a leave positive effect on birthrate.

As we have seen, studies of women's earnings/wages concerning childcare in Korea began at the end of the 2000's. Still, any attempt to determine the earnings effects of women having children, as in this study, remains a novelty in Korea. Further, a comparison of human capital and women's earnings in Korea with that in Germany according to the same empirical framework constitutes a new contribution to the research on women. This study plans, at the beginning of estimation, to give a general presentation of differences in earnings concerning children, family characteristics and education in Korea and in Germany using the quantile regression method. The intensive analysis of the earnings effect of women having children is then carried out by mean of the regression discontinuity design. Low birthrate, which has been recognized as an important socio-economic problem in Germany, is also becoming a serious problem in Korea. What differences/similarities does having children have on a woman's career in both countries? This study offers a chance to look at this problem from a new angle.

Following this introduction, I move on to a consideration of the differences in education, employment and fertility rate of women in Korea and in Germany, which are mainly based on the data collected by the OECD. Thereafter, I will introduce models and methods in the third Section. This Section outlines the model developed by Mincer and Polachek (1974), the quantile regression and regression discontinuity design techniques. In the next Section, the data and variables are explained, and descriptive statistics are presented. The data used in this study are taken from the Korean Labor and Income Panel Study (KLIPS) and German Socio-Economic Panel (GSOEP). In the fifth Section, estimation results are obtained, and the following conclusion finalizes this study.

3.2. Education, Employment and Fertility Rate of Women in Korea and Germany

Korean Women

Observing the participation rates of women in the higher level of education, the female entry rate for tertiary education 'type B'[22] (junior college level of education) averages 53% based on each year of relevant age, while the value is 47% for males in Korea. In the OECD on average, the entry rate for tertiary education 'type B' is 17% for females and 13% for males. The entry rate for tertiary education 'type A' (university level of education) amounts to 59% for females and 63% for males, while these values are 63% for females and 50% for males in the OECD on average (OECD (2009a)). The male entry rate to university is higher than the female one in Korea and vice versa in the OCED on average. In 2009, the female entry rate to tertiary educational institutions – junior college and university – for all senior high school graduates, however, exceeded the male entry rate for the first time (Korean Statistical Information Service (2010), pp. 107).

The employment rate of Korean women between the ages of 25 and 64 is relatively low at 58.3% compared to an OECD average of 64.9% in 2007. Compared to the average for the OECD (79.5%), women with tertiary education (i.e. junior college and university degree) have much lower employment rates

22 Tertiary education programs in 'type B' are generally more practical/ technical/ occupationally specific than 'type A' programs. Qualifications in category B are typically shorter than those in A and focus on occupationally-specific skills geared for direct entry into the labor market, although some theoretical foundations may be covered in the respective programs. OECD includes the following institutions in this category in Korea: Junior college, Polytechnic college, Technical college, Third military academy, National college of taxation, Open university, Polytechnic university, Military academy, National college of police, University of education, (Military) Nursing academy, Naval academy, Air force academy, School of national securities, Graduate school-special, Graduate school-professional. See OECD (1999).

(61.2%), while women with an education level below upper secondary have much higher employment rates (58%) compared to the OECD on average (47.8%) in 2007. Korea is thus only an exceptional case among all OECD countries, where women who attain university level education generally have a higher participation rate in the labor market than women with lower educational levels. Average annual earnings for females as a percentage of males' aged between 25 and 64 are 48% for women with an education level below upper secondary, 47% for women with an upper secondary educational level and 65% for women with a tertiary education degree in 2003. One third of working women have only a temporary work contract and only 8% of working women have a supervisory responsibility, which leads to a high gender earnings gap in Korea (OECD (2009a)).

Figure 3.1: Public Spending for Family in Cash, Services and Tax Measures (Percentage of GDP, in 2003)

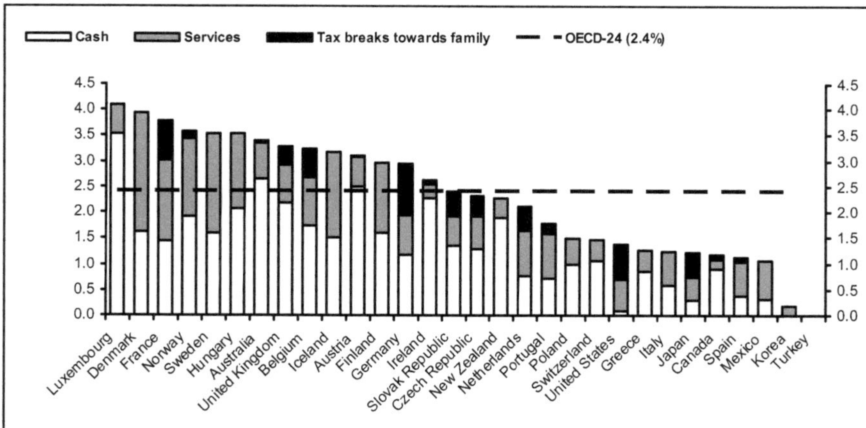

Source *OECD (2007).*

Concerning public support exclusively for families such as child payments and allowances, parental leave benefits and childcare support, Korea achieved the lowest position amongst all the OECD countries. Public support in monetary payments stands at 0.175% (2.4% for 24 OECD countries[23] on average) of GDP: 0.008% is due to direct fiscal support and 0.167% to public services. There is no tax break for families in Korea (see Figure 3.1). Under such social circumstances, it is hardly possible for women to balance jobs and childcare

23 OECD-24 on this Figure excludes Greece, Hungary, Luxembourg, Poland, Switzerland and Turkey where tax spending data are not available.

whilst supporting their family without, for example, having their mother or mother in law helping in caring for any children. Firms in Korea expect that women will usually leave regular employment around childbirth. On the other hand, Koreans have a fixed notion that the education of children is not only important for the children's future, but also for the family honor. So, attending to the schooling of children could be considered more important for women in Korea compared to other countries in different cultural circumstances. This could also lead to a weak attachment to work and a low employment rate for more highly educated women.

The OECD (2009) reports the lowest fertility rate for Korean women in a group of 30 OECD countries. Women aged between 15 and 49 had just 1.08 babies on average in 2005. The fertility rate in Korea has dramatically decreased since 1960 when about 6 children were born per woman. Since 2002, the total fertility rates in Korea have remained below the level of 1.2 babies (see Figure 3.2).

German Women

Concerning the educational participation of German women, the entry rate to educational institutes of tertiary 'type B'[24] (An educational level comparable to German *Meisterschule*) averages 16% based on each year of relevant age for women and 10% for men. These are slightly lower than the OECD average for 2007. The entry rate to tertiary education 'type A' (*Fachhochschule* and university) is 35% for females and 34% for males, which is a relatively low for most OECD countries (OECD (2009a)).

German women between the ages of 25 and 64 who have attained a tertiary level education had an average participation rate of 80.6% in the labor market in 2007. This is slightly higher than the OECD average. Women with an education level below upper secondary and an education level of upper secondary have 47.3% and 68% employment rates (i.e. 47.8% and 67.2% OECD on average). Women who have attained a university level education have higher participation rates in the labor market than women with lower levels. The employment rates for females are high at 67.3% for all educational levels compared to an OECD average of 64.9% in 2007. Average annual earnings for German women as a percentage of earnings of men between the ages of 25 and 64 were 55% for an edu-

24 German tertiary education programs in type B are Health sector schools, Trade and technical schools, Bavaria specialized academy, Schleswig-Holstein vocational academies and Baden-Wuerttemberg vocational academies. See OECD (1999).

cation level below upper secondary, 59% for an education level of upper secondary and 59% for a tertiary education degree in 2007 (OECD (2009a)).

Figure 3.2: Total Fertility Rates from 1970 to 2006

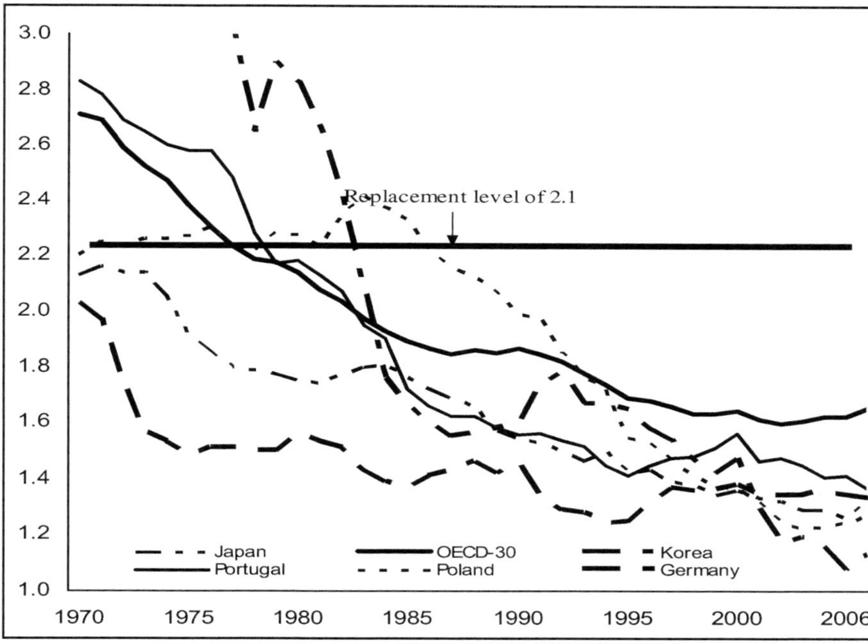

Source *OECD (2009b).*

Concerning public support exclusively for families, monetary public support is equal to 1.2% of GDP, with public services at a level of 0.77% and tax breaks towards the family at 1.0%. This is equal to a total of 3.0% of GDP in 2003 (OECD (2007)). Total public spending for families with children is relatively high in Germany compared to other OECD countries. But spending on services is relatively low and is mostly allocated to supporting pre-school (kindergarten) children aged 3-6: Formal childcare for children younger than three years old remains at 9% in Germany, while this has an average OECD value of 23%. 80% of children aged 3 to 5 are cared for in kindergarten (i.e. 73% OECD on average) but usually on a part-time basis, as out-of-school-hours care services are not often available. The fertility rate of Germany (1.34 babies in 2005) is higher than Korea's. But many German women remain childless: 36% of women aged 41- 44 have no children, while this is around 20% or less in most OECD countries. German fertility rates have decreased relatively slowly compared to Korea:

In 1960, a woman had 2.37 babies in Germany, while this was 6 children in Korea. However, the fertility rate has remained under the replacement rate (2.1 babies) for quite a long period of time. This country has suffered from low fertility since the 1970's (see Figure 3.2).

3.3. Model and Methods

Mincer and Polachek's model (1974) seeks to explain the earnings of women according to their individual skills and work histories, considering not only experience at work, but also discontinuities of working experience due to childcare. According to the model, earnings can be expressed in the following way:

$$\ln E = a_0 + \alpha s + \beta(e_1 + e_2 + e_3),$$

where $\ln E$ is equal to log gross earning, s represents skills such as school education and training, e_1 working experience before marriage, e_2 home time, which is time for childbirth and child rearing and e_3 working experience after home time. α stands for returns to schooling and β returns to working experience. a_0 is intercept (Mincer and Polachek (1974), p.87).

Based on this model, the primary purpose of this study is to estimate the effect on women's earnings of having children, which constitutes the main discontinuity in a women's working career. In the estimation of human capital effects on women's earnings, age and tenure are employed as measures for women's working experience. Child number and marital status are used as proxies of home time, which could serve to distinguish the effects of different work histories between women. Household head and single mother index, which could strengthen job attachment and effort according to Becker (1985), control for their influence on the working experience and earnings of women. Apart from tasks designed to study the effects of the different working histories of women, the estimation obtained by using Mincer and Polachek's earnings function still suffers from the general problems common to empirical human capital study: Measurement error, omitted ability bias and selection bias concerning schooling. Even studies by Ashenfelter and Zimmerman (1993, 1997) demonstrated that the omitted ability bias and the measurement error bias due to schooling canceled each other out using the OLS, though the study could not totally remove the biases. The selection bias attributable to schooling remains. As a comparative study, with more limits on the availability of data and empirical techniques[25], this study focuses on demonstrating relative differences in educational effects on earnings between both countries in the presentation of human capital effect on earnings.

The methods, which this study utilizes, are the quantile regression and the regression discontinuity design (RD design). The quantile regression method gives a comprehensive presentation of the earnings effects attributable to a school education and family characteristics. The regression discontinuity design focuses on estimating the effect that having one or two children has on earnings in two different experience statuses: This study treats women aged between 30 and 40 as a less experienced group suffering more strongly from career intermittence, and women aged between 40 and 55 as a more experienced group suffering less from career intermittence.

Quantile Regression A prominent advantage of quantile regression is that this method enables us to analyze several different regression curves along with the various distributional percentile points of log earnings unlike the least squares regression (Koenker and Bassett (1978); Koenker and Hallock (2001); Machado and Mata (2001); and Koenker (2005)). The quantile regression technique reveals the earnings differentials between and within levels of covariates. In a comparative study, this advantage especially expands the comparability manifold. In the study, the quantile regression method allows us to measure earnings differentials attributable to the variables of school education and family characteristics throughout the whole distribution in Korea compared to those in Germany as well as to look at the effect of workers' heterogeneity on the earnings between both countries. This study presents inter/intra-earnings differentials in educational levels and years, numbers of children, marital status, household head and single mother indicators at five quantiles of the log earnings distribution – at the 10th quantile, at the 25th quantile, at the median, at the 75th quantile and at the 90th quantile.

Regression Discontinuity Design Thistlethwaite and Campbell (1960) originally introduced the regression-discontinuity design. This method tries to remove the

25 As countermeasures to these biases, using a proxy variable of ability, an instrument variable of schooling, employing the fixed effects method, Heckman 2 step method (Heckman (1979)) and the control function method (Garen (1984)) are proposed, as long as appropriate variables and techniques are available.

self-selection bias between treated and control groups through its special assignment technique.

The self-selection bias has been pointed out as a problem in estimating the treatment effect of a binary explanatory variable (i.e. having children or not in this study) using the OLS. When estimating the following earnings function, $earnings_i = a_0 + \beta \cdot treatment_i + u_i$, where *treatment* is indicated by one, for example, when some women indexed with *i* have children and zero when some women have no children. In this case, what we really want to estimate using this treatment variable is the difference between the earnings effect of having children and not having children on the same women in the randomly selected sample. However, we actually estimate the earnings difference between some women having children and some other women having no children. But women having children could have certain social and individual characteristics, which differ from women without children. So, this could cause self-selection bias and β suffers from the dependence between $treatment_i$ and error term u_i. Rosenbaum and Rubin (1983) demonstrated that the treatment assignment can assure 'the ignorability of treatment' given sufficient covariates *x* which could balance any self-selection bias. In this case, adjustment for observed covariates *x* is sufficient to produce consistent estimates of treatment variables, on which the regression discontinuity design method is also founded.

RD design employs a special treatment assignment technique which depends on the value of an observed continuous variable (i.e. tenure in this study) related to a given cutoff point. This continuous variable is an element of the covariates *x*. Following the treatment assignment technique of the RD design, this study selects women who are below the given cutoff point of tenure for the treatment group (*T* =1), while women above the cutoff point are selected for the control group (*T*=0). If this assignment arises through a known and measured deterministic decision rule, the model belongs to the sharp RD design (Trochim (1984)) as is the case in this study: $T_i = 1\,\{tenure_i \leq \overline{tenure}\}$, where *i* is the index for women and $\overline{tenure}$ a cutoff point (i.e. sample mean of tenures in this study). Concerning pure randomization and a strong ignorability condition, however, the dependence of treatment assignment on an observed continuous variable has a particular problem. In order to identify treatment effects consistently, the RD design limits the sample to a very small interval around the cutoff point of the continuous variable. In this study, the samples are limited to the range of +/- one year of tenure at the cutoff point. According to the idea of the RD design, this condition could be very similar to a randomized experiment (a tie-breaking experiment). Women having approximately the same tenure above and below the cutoff score could have similar average earnings.

Concerning discontinuity, if the treatment effect (i.e. effect of having children) differs from the earnings effect of the control group (i.e. effect of having no children), we could find a break at the cutoff point along the tenure-earnings function. We can identify a treatment effect β employing the RD design in the following way (Klaauw (2002), pp.1258):

$$\begin{aligned}&\lim_{ten\downarrow \overline{ten}} E[earnings \mid tenure] - \lim_{ten\uparrow \overline{ten}} E[earnings \mid tenure] \\ &= \beta + \lim_{ten\downarrow \overline{ten}} E[u \mid tenure] - \lim_{ten\uparrow \overline{ten}} E[u \mid tenure]\end{aligned} \qquad (1),$$

where *ten* stands for tenure and $\overline{ten}$ cutoff point in this function. When the treatment effect is allowed to vary randomly across individual, then equation (1) will determine the average treatment effect of those $E[\beta_i \mid \overline{tenure}]$ at the margin. In the case where treatment effects vary in a deterministic fashion with tenure, equation (1) will identify the local treatment effect $\beta(\overline{tenure})$ at $\overline{tenure}$.

There is a further problem with RD design: The treatment status of an individual *i* (i.e. a woman has children or not) could be related to the observed characteristic of tenure which is itself related to the outcome variable of earnings. So, the dependency between the assignment variable $tenure_i$ and the error term u_i, $T = T(tenure_i) = 1\{tenure_i \leq \overline{tenure}\}$, might lead to a biased and inconsistent OLS estimate of β. As a countermeasure, one includes a control function proposed by Heckman and Robb (1985), $C(tenure_i, treatment_i)$, which is specified using semi-parametric, local, non-parametric or Kernel regression estimators.

The formulation of the earnings function used in this study, according to the RD design, is: $earnings_i = a_0 + \beta \cdot treatment_i + C(tenure_i, treatment_i) + \phi x_i' + u_i$, where the control function, $C(tenure_i, treatment_i)$, is specified by local polynomial form in this study. ϕ is a coefficient vector for the other control variables. This study estimates the earnings effects associated with having a single child, two children or combines the effect of having one to two children compared to no children considering the average tenure of female employees using the following function, excluding index *i*:

$$\ln earnings = a_0 + \beta_1 treat + \beta_2 ten + \beta_3 ten^2 + \beta_4 treat \cdot ten + \beta_5 treat \cdot ten^2 + X\phi + u.$$

Where, *treat* represents treatment variable, *ten* equals tenure variable and *X* includes age, age squared and dummy variables concerning marital status, household head, single mother, education, job training, public firm, overtime, part time, firm size, family residence, occupation, industry and year. If *treat* equals one, $\ln earnings = a_0 + \beta_1 treat + (\beta_2 + \beta_4)ten + (\beta_3 + \beta_5)ten^2 + X\phi + u$. If *treat* equals zero, $\ln earnings = a_0 + \beta_2 ten + \beta_3 ten^2 + X\phi + u$.

3.4. Data and Descriptive Statistics

The data, which is used in this study, is taken from the Korean Labor and Income Panel Study (KLIPS) and the German Socio-Economic Panel (GSOEP) between 2002 and 2007. The KLIPS, launched in 1998, is a longitudinal survey of the labor market and income activities of households and individuals in 7 cities and 8 urban areas. As far as design and data-management are concerned, the panel study is comparable to the GSOEP, which was a role model for KLIPS. The panel takes into consideration a variety of variables concerning social and economic backgrounds. This survey has served as a data source for the microeconomic analysis of labor market activities and transitions, thereby contributing to the development and evaluation of labor market policies. The GSOEP, launched in 1984, is published by the German Institute for Economic Research (DIW). The data are taken from 8 main areas in Germany and consist of 8 samples: *Sample A* to *Sample H*. *Sample A* covers 'Residents in West Germany (the Federal Republic of Germany)' and *Sample B* 'Foreigners in West Germany.' *Sample C* surveys 'German Residents in East Germany (the German Democratic Republic).' *Sample D* for 'Immigrants' was launched in 1994/95. Sample E 'Refreshment' was selected independently of the ongoing panel (subsamples A through D). Since 2002, there is also *Sample F* 'Innovation' and *Sample G* 'Oversampling of High Income.' In 2006, *Sample H* 'Refreshment sample' was again initiated with 1,506 households selected from the total population. The data includes a large number of socioeconomic variables concerning population and demography, education, training, qualification, labor market and occupational dynamics, earnings, income, social security, housing, health, household production, basic orientation and satisfaction with life in general, and certain aspects of life etc.

This study excludes *Sample G* due to oversampling from the high income group and analyzes only those individuals with a West German educational background in order to establish a comparable estimation of the different earnings effects of schooling between Korea and Germany. Further, in order to reduce the selection bias at the beginning and the end of vocational life, I use the data concerning female employees aged between 30 and 55 taken from the GSOEP and the KLIPS. This study considers 3,066 observations in Korea and 8,260 observations in Germany for women aged between 30 and 55 utilizing the following variables: Gross yearly labor earnings of individuals, age and tenure, marital status, single mother, household head, children number and educational levels/years, job training, public firm, overtime, part time, firm size, family residence, occupation and industry.

The main variables utilized in this study are the gross yearly labor earnings of individuals, age, tenure and dummies for marital status, single mother, household head, children number and educational levels/years. The dependent variable is the log of yearly gross labor earnings for the last calendar year under consideration, which averages 9.17 in Korea and 9.78 in Germany. These average gross earnings for women, in the year under consideration, are about 11,544.60 Euro in Korea and 22,536.86 Euro in Germany. Choosing the average yearly gross earnings as the dependent variable has an advantage, when compared with current mean wages and salaries from principal employment: The gross yearly labor earnings of individuals includes not only wages and salaries from the main job, but also wages and salaries from any additional job. Moreover, these earnings include not only labor earnings for employees, but also earnings for the self-employed. This enables us to reduce sample selection bias.

In the sample, the average age for women amounts to about 41 years in Korea and about 42 years in Germany. Tenure on average is 4.2 years in Korea and 10.2 years in Germany. Among women aged between 30 und 55, 8.22% are single, 80.1% married, 6.03% widowed, 4.11% divorced and 1.63% separated in Korea, and 18.20% are single, 66.04% married, 1.42% widowed, 11.76% divorced and 2.59% separated in Germany. Further, 15.20% of these women are household heads and 3.98% are single mothers in Korea, while 43.33% of them are household heads and 10.40% single mothers in Germany. In the sample, 40.04% of women have no children, 20.22% one child, 33.76% two children, 3.91% three children and 0.07% four children in Korea, while 52.68% have no children, 24.08% one child, 18.80% two children, 3.72% three children, 0.67% four children and 0.06% five children in Germany.

Educational variables comprise educational levels and educational years. Educational levels are classified into 5 categories in the following way:

1. The lowest level is composed of individuals having an education level lower than junior high school graduation in Korea and an education level lower than completion of general elementary schooling (*Realschule* or *Hauptschule*) in Germany.
2. The next lowest educational level includes a graduation from junior high school in Korea and a completion of general elementary schooling (*Real-* or *Hauptschulabschluss*) in Germany.
3. Senior high school graduation in Korea is compared with qualification for college or university and/or vocational qualification (*Berufsabschluss* and/or *(Fach-)Hochschulreife*) in Germany. This type of education is classified as the third level.
4. Graduation from junior college in Korea is comparable to the higher vocational qualification (*Abschluss von Tehnikschule/Meisterschule* etc.) in Germany, which are together assigned to the fourth level.
5. The highest educational level: A university degree or higher in Korea is compared with a graduation from university focusing on vocational practice (*Fachhochschule*) or higher degrees in Germany.

In the datasets, 12.33% of women have an education level lower than junior high school graduation in Korea, 19.60% graduated from junior high school, 45.60% from senior high school, 8.94% completed junior college and 13.54% have university degrees or further qualifications in Korea. In Germany, 0.59% of women have an education level lower than *Real-* or *Hauptschulabschluss*, 11.91% graduated from a *Realschule* or a *Hauptschule*, 61.92% have a *Berufsabschluss* and/or a *(Fach-)Hochschulreife*, 8.78% have a higher vocational qualification and 16.79% graduated *Fachhochschule* or more advanced educational institutions.

In order to make a better comparison between the earnings effects of education in both countries, this study additionally estimates the earnings effects of educational years as a quantitative measure of education. The duration of any education, measured in years, is divided into 7 categories in KLIPS and into 16 categories in GSOEP. Among these categories, 9, 12, 14, 16 and 18+ years of education are common categories in both countries, which are compared each other. This years used in this study are not actual years of education completed for each woman, but the standardized durations of each educational program. Using variable educational years, instead of education levels, this study makes use of 3,065 observations in Korea and 8,217 observations in Germany in both datasets. In Korea, 19.61% of the Individuals considered in these observations have 9 years education, 45.61% 12 years, 8.94% 14 years, 11.45% 16 years, and 2.06% of individuals have completed 18+ years of education. In Germany, 7.33% of observations concern individuals with 9 years of education, 12.75% 12 years, 1.61% 14 years, 3.63% 16 years, and 10.13% of individuals have completed 18+ years of education. Descriptive statistics of other control variables are presented in the appendix (see Table 3.1 and Table 3.2).

In order to estimate the effect having children has on earnings by means of the regression discontinuity design, this study divides observations into two age groups: Women aged between 30 and 40, and women aged between 40 and 55. For the former group, age and tenure average 34.7 and 4.01 years in Korea and 34.5 and 6.99 years in Germany. For the latter group, the age is 46.2 in Korea and 46.7 in Germany on average. Tenure amounts to 4.3 years in Korea and 12.4 years in Germany. After treatment assignment according to the RD design technique, this study used the restricted datasets concerning the variable of tenure: I limited the data to a range of +/- one year of average tenures. Descriptive statistics for the main variables used in this part of the study – earnings, age, tenure and treatment – are presented in the appendix (see Table 3.3 and Table 3.4).

3.5. Estimation Results

3.5.1. Effects of Human Capital on Earnings for Women using the Quantile Regression Method

Age and Tenure Table 3.5 and Table 3.6 present the earnings effects of age and age squared and tenure and tenure squared in Korea and Germany. Except at the 25th quantile, age does not demonstrate significant effects on earnings in Korea, while age has high significant effects on women's earnings in all models in Germany. Tenure generally leads to significant earnings effects in both countries. Analyzing age-earnings profiles, the concave curves illustrate the greater effects of age on earnings in Germany compared to Korea by means of OLS and at all quantile points using the quantile regression method. Earning effects attributable to age attain their highest levels about the ages of 37 at the 25th quantile in Korea, while the effects are highest around the ages of 45 (OLS) and 46 (median) in Germany. Most tenure-earnings profiles indicate a stronger effect on earnings in Korea than in Germany by OLS and at all quantile points. Tenure effects on earnings attain their greatest levels between the tenure of 44 and 45 by the OLS, and at about the tenure of 46 at the median in Korea. The greatest tenure effects are evident at tenure of approximately 27 by the OLS and at tenure of about 25 at the median in Germany (see Figures 3.5 to 3.8).

Number of Children Table 3.7 and 3.8 illustrate the effect of having children on women's earnings in Korea and in Germany using dummy variables indicated by numbers of children. The base variable is the variable for having no children. According to Table 3.7, we can see that the fact of having children, for the most part, does not result in significant earnings effects in Korea at the usual significance level. Trends can be noticed in the result, such as: The positive effects associated with having a single child or 4 children and the negative effects of having 2 or 3 children. According to the OLS estimation, most results are statistically insignificant. Only women who have three children earn, with all other factors fixed, about 8% less than women who have no children in Korea at the 10% of significance level. At the median, earnings effects attributable to numbers of children do not present any significant results. At the highest quantile (90th), having children has purely negative effects on earnings. Analyzing within earnings differentials associated with the number of children, women having one child earn significantly 5.4% more at the 25th quantile compared to the base variable. Having two children leads to significantly negative earnings effects at the 10th and 90th quantile (i.e. -8% and -6.5% respectively). Women who have three children earn -10% and -17% less at the 25th and 90th quantile respectively, when compared with the base variable. The variable associated with having 4 children shows significantly positive results at the 10th, 25th and 75th quantile (i.e. 62%, 48% and 31% in earnings respectively).

In Germany, variables concerning the number of children mostly present negative earnings effects up to 5% significance level. Using the OLS, having a child leads to a -11% drop in wages, having two children to a -19% fall, three children -21%, four children -19% and five children a -42% earnings decrease compared with women having no children. At the median, these present -11%, -17.5%, -18%, -12% and about -25% respectively. Observing intra-levels differences, the earnings effects of having a child fluctuate between -6% (90th) and -14% (25th), having two children from -11% (90th) to -24% (25th), having three children -18% (50th) to -23% (25th), having four children -12% (50th) to -23% (10th) and having five children from -25% (50th) to -55.4% (25th). By trend, we witness a larger reduction in earnings due to children in the lower quantiles when compared to the higher quantiles (see Table 3.8). Comparing the estimation result related to the number of children in a household in Korea with that of Germany, this study concludes that the effects on earnings of having children are much more significant in Germany than in Korea (see Figures 3.9 and 3.10).

Marital Status and Family Responsibilities Table 3.9 and Table 3.10 illustrate the earnings effects of marital status, household head and single mother in Korea and in Germany. The base variable of marital status is set equal to the state of a single woman. Women who have family responsibilities as household heads are compared with women who are not household heads. Single mothers are compared to other women. In Korea, marital status seldom results in a significant effect. Focusing on particularly significant results, married women earn 11% more than single women at 90th quantile. Separated women earn 15% more by the OLS, 20.5% and 34% more at the 75th and the 90th quantile compared with single women. Women who have family responsibilities as household heads significantly earn more than other women in Korea: 9% (10th, 25th and OLS) and 7% at the median. Family responsibility as single mothers results in an about 10% higher earnings effect by means of the OLS and at the 25th and 90th quantile estimation with a significance level of around 10% compared with other women. At the lowest quantile (10th), the earnings effect of being a single mother reaches its highest level (22%) at around 5% of significance level.

In Germany, married women earn significantly less than single women: -6% by the OLS and -4.6% at the median. These fluctuate between -4% (75th) and -9.3% (10th). The effects are stronger at the lower quantiles. Widowed women earn significantly less at the median than single women (-8%), while this is insignificant according to the OLS. At the 70th and 90th quantile, these effects attain a value of -8.4% and -8% respectively up to a 10% significance level. Variables for divorced and separated women do not yield any significant estimates up to 10% significance level. Women who have family responsibilities as household heads earn 2.7% (OLS) and 3% (25th) more than other women up to 10% significance level, while this is insignificant at the median. At the highest quantile (90th), female household heads are seen to result in a 4% significant earnings effect. Single mothers with family responsibilities earn more than other women: 17% (OLS), 23.6% (10th), 22% (25th), 18% (50th), 17% (75th) and 7.2% (90th) up to 1% of significance level. Compared to Korean women, marriage leads to a much more significant earnings decrease in Germany. Family responsibility as a household head or a single mother results in significant positive earnings effects in both countries. In particular, the effects of being a single mother in Germany are noticeably high (see Figures 3.11 to 3.13).

Educational Level Table 3.11 and Table 3.12 illustrate the earnings effects attributable to various levels of school education. The base variable for educational level is senior high school graduation in Korea and the attainment of a vocational qualification and/or qualification for college or university (Berufsabschluss and/or (Fach-)Hochschulreife) in Germany. The estimation result in general gives us highly significant effects on earnings in both countries. An exceptional case is the estimates obtained when considering the variable of junior high school graduation in Korea, which are insignificant in all models. Table 3.11 and Figure 3.3 illustrate the result that women who have a lower educational level than junior high school graduation earn -9.5% less than women who graduated senior high school by means of the OLS. Graduation from junior college leads to 8% higher earnings when compared with the base group. Women acquiring at least a university degree earn 40% more than senior high school graduates. At the median, an education level lower than junior high graduation results in, on average, a -10.4% earnings decrease compared to the base group. Junior college graduation has an 11% positive earnings effect and university graduation a 44% positive effect. Turning our attention to earnings differentials within educational levels, an educational level lower than junior high school graduation leads to a -8.4% (25th) to -11% (90th) significant earnings decrease compared to the base group. Earnings differentials when considering junior college graduation fluctuate between 9.5% (90th) and 11% (25th). Women with university degrees earn 35% (10th) to 47% (75th) more than women who graduated senior high school at the 1% significance level.

Table 3.12 and Figure 3.4 illustrate the estimation results for German women. By means of the OLS, women who have a lower educational level than *Real-* or *Hauptschulabschluss* earn -65% less than women with a *Berufsabschluss* and /or a *(Fach-)Hochschulreife*. *Real-* or *Hauptschulabschluss* presents -8.4% lower earnings than the base group. Women who have acquired a higher vocational qualification, such as completion of a course at a *Technik/Meisterschule*, earn 7.8% more than the base group. Women with a degree from a *Fachhochschule* and/or higher educational level earn 25% more than women with a *Berufsabschluss* and/or *(Fach-)Hochschulreife*. At the median, an education level lower than *Real-* or *Hauptschulabschluss* results in an approximate -55% earnings decrease compared to the base group. *Real-* or *Hauptschulabschluss* result in a -4.8% earnings decrease. The higher vocational qualification leads to a 11% positive effect and a *Fachhochschulabschluss+* to a 22% positive effect on earnings compared with the base variable. As we have seen, earnings differentials between the lowest and highest educational levels are high, especially at the 10th and 25th quantiles.

Figure 3.3: Effects of Educational Levels on Women's Earnings in Korea

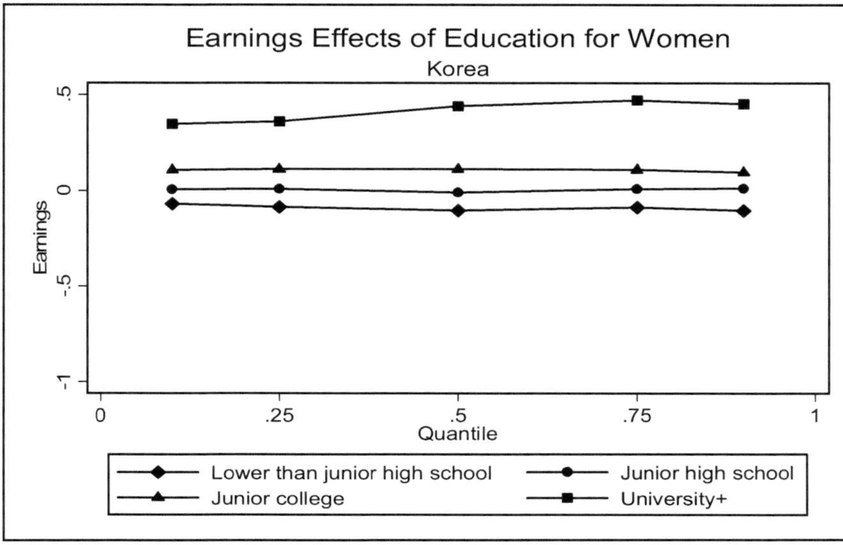

Note *Base variable for education level: Senior high school.*

Analyzing intra-educational level earnings differentials, an educational level lower than *Real-* or *Hauptschulabschluss* leads to -84% (25th) to -41% (90th) significant earnings decreases compared to the base group. A *Real-* or *Hauptschulabschluss* leads to -13% (10th) to -4% (75th) earnings decreases. The effects of higher vocational qualifications on earnings fluctuate between 4.6% (75th) and 17% (25th). The earnings effects of attaining a *Fachhochschulabschluss+* result in gains in earnings of 22% (50th) to 28% (25th). Women at the lowest educational level and at the lowest quantile point experience noticeably high intra /inter-educational level earnings differentials in Germany.

Compared to Korea, the between and within educational level earnings differentials are much higher in Germany. Women with low levels of education suffer from especially low earnings levels in Germany. However, the earnings differentials between an educational level equivalent to a junior college degree in Korea and its German counterpart – the higher vocational qualification – are very similar. A university degree is more profitable for Korean women than German women at all quantile points and at mean using OLS.

Figure 3.4: Effects of Educational Levels on Women's Earnings in Germany

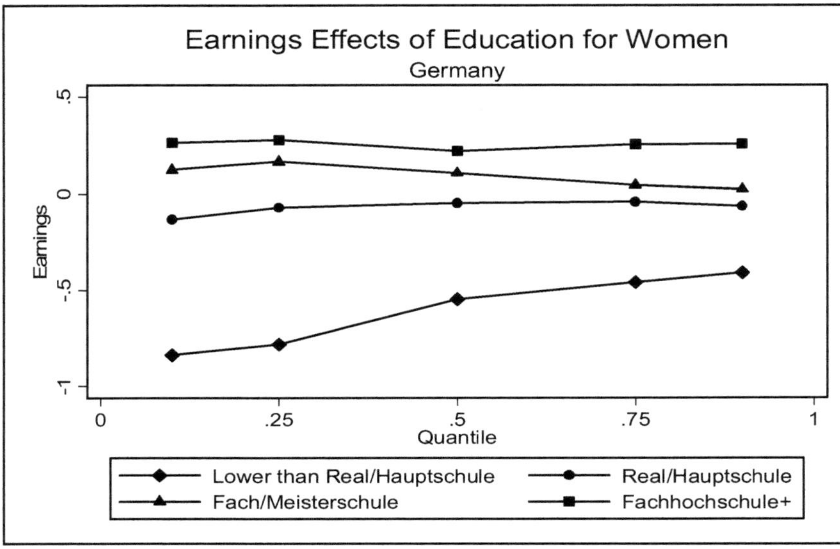

Note *Base variable for education level: Berufsabschluss and/or (Fach-)Hochschulreife.*

Educational Year Table 3.13 and Table 3.14 illustrate the effects that the number of years spent in education have on earnings in Korea and in Germany. The base variable is 12 years of education, which represents a completed senior high school education in Korea, and the completion of specialized vocational high school (2 years *Fachoberschule*) and specialized vocational high school necessary for university entrance or starting on a skilled occupation (*Berufsfachschule*) in Germany (OECD (1999)). Table 3.13 illustrates the effects of years spent in education on women's earnings in Korea. The estimation result demonstrates that more educational years lead to higher earnings. Women possessing no education earn 16.6% (OLS) and 22% (median) less than women who have been educated for 12 years. The variable for 9 years education does not produce any significant results in all models. 14 years of education which is equivalent to junior college graduation in Korea leads to an earnings increase of 8.3% (OLS) and 12.7% (median) compared to the base group. 16 years of education, i.e. bachelor's degree at university, leads to a 38% (OLS) and a 42% (median) earnings increase compared with the base group. 18+ years of education which is equivalent to a master's and doctoral degree has a 52% (OLS) and a 67% (median) positive effect on earnings. Focusing on estimates at the usual significance level, within level earnings differentials vary -18% to -22% for zero years of education, 10% to 13% for 14 years of education, 35% to 42% for 16 years of education and 52% to 67% for 18+ years of education by the quantile regression.

Table 3.14 illustrates the effect of years spent in education on women's earnings in Germany. More years spent in education in general lead to higher earnings. However, there are some exceptional cases: Using the OLS, 11 years of education has a significantly weaker earnings effect (-9.8%) than 10.5 years of education (-6.6%), and again the earnings effect attributable to 13 years of education (-13%) is significantly weaker than that of 11 years of education etc. The lowest educational duration (7 years) is correlated with a strong earnings loss compared to the base group: -69% (OLS) and -60% (median) at the 1% significance level. Women with 9 years of education earn -14% (OLS) and -13% (median) less than the base group. The variable for 14 years of education does not give any significant results in all models. 16 years of education leads to an earnings increase of 16% (OLS) and 15% (median) compared to women educated for 12 years. Women educated for 18 years and longer earn 35% (OLS) and 27% (median) more than the base group. Earnings differentials within years of education vary -49% (90th) to -87% (10th) for 7 years of education, -10.6% (75th) to -28% (10th) for 9 years, 13% (10th) to 20% (75th) for 16 years and 27% (50th) to 49% (10th) for 18+ years of education.

Compared to Korea, women with the least number of years spent in education suffer more severely from lower earnings in Germany. Earnings gains due to longer educational durations, such as 16 years and 18 years+ of education, are smaller in Germany than in Korea. At the lowest quantile point, between and within earnings differentials concerning years of education are higher in Germany than in Korea (see Figures 3.14 and 3.15).

3.5.2. Discontinuity of Earnings due to Children using the RD Design

Women aged between 30 and 55 Table 3.15 and Table 3.16 illustrate the treatment effect of having either one or two children, or the combined effect having one or two children has on earnings compared to having no children for women aged between 30 and 55. The cutoff points between treatment and control groups are 4 years of tenure in Korea and 10 years in Germany, which are approximately average tenures for each sample (i.e. 4.21 and 10.21 years respectively). Below the cutoff points, women have children, while women have no children above the cutoff points. The former are the treatment groups and the latter control groups. According to the idea of the regression discontinuity design, this study analyzes the data within this very small interval around the cutoff point. These are +/- one year from cutoff points: 3 to 5 years of tenure in Korea and 9 to 11 years of tenure in Germany.

In Table 3.15, we can see the estimation result for the Korean dataset. As a reference, the table gives us the result estimated for all tenure excluding the bandwidth of tenure 3 to 5 years. Estimation models are first divided into 3 groups: women having one child, having two children or having one or two children. Each group is subdivided into 2 estimation models, which use local linear control functions of tenure and treatment or which include their squared form. We seek to interpret the estimation result obtained by this model by estimating the earnings effect associated with having one child and employing a local linear control function with 3 to 5 years of tenure bandwidth: The treatment variable of having one child has approximately a 13.6% insignificant earnings effect compared to the control group who have no child. The tenure earning function demonstrates a positive significant effect with value of 0.28 on the earnings and interaction term which controls dependence between the assignment variable $tenure_i$ and error term u_i shows a negative trend, and has an negative earnings effect with value of -0.26 at about 15% significance level. In Table 3.15, this study does not find a statistically significant treatment effect limiting the data within the tenure bandwidth. Removing the tenure bandwidth, we find significant effects associated with having one child compared to having no children: 22% and 14% using a local linear function and including the squared form respectively. However, the estimates could be contaminated by tenure-earnings profiles which are influenced by self-selection bias.

Table 3.16 gives us the result for the German dataset, which also does not lead to any significant results. Around the average tenure (i.e. 10 years), there are no significant effects for treatment variables in all models. Their control functions are also statistically insignificant. Without the 9 to 11 years bandwidth of tenure, having two children results in -12.5% and -22% significant negative earnings effects using linear and squared functions of tenure and treatment respectively, and having 1 to 2 children -6.4% and -12.6% significant earnings effects. However, the results could be biased due to the selection bias in the functions of tenure and earnings.

Women aged between 30 and 40 Table 3.17 and Table 3.18 tabulate the effects of having children on earnings around the cutoff point (i.e. 4 years in Korea and 7 years in Germany) for women aged between 30 and 40. The sample mean of their tenure amounts to 4.01 years in Korea and 6.99 years in Germany. Table 3.17 tabulates the estimation results in Korea. The treatment effects do not have any significance when considering a tenure bandwidth of between 3 and 5 years in Korea. In Germany, women aged between 30 and 40 give significant results when considering a tenure bandwidth of 6 to 8 years. Using the control function in its squared form, having one child results in -49% (around 15% of significance level), having two children -70% and having one to two children -65% negative earnings effects compared to having no children up to 5% of significance level. In case of having one to two children, tenure and tenure squared have a value of -.97 (i.e. β_2 in the model of this study.) and a value of 22.7 (β_3) earnings effects at the 5% significance level respectively. The interaction term of tenure and treatment and its square have values of 21.8 (β_4) and -.97 (β_5) and lead to earnings effects at the 10% and 5% significance level respectively. There is thus a break at the cutoff point between the tenure-earnings function, $earnings = (\beta_2 + \beta_4)ten + (\beta_3 + \beta_5)ten^2$, for women with one or two children and the tenure-earnings function, $earnings = \beta_2 ten + \beta_3 ten^2$, for women with no children could be significantly confirmed. These results could assure discontinuity of earnings between the treatment and control groups due to children using RD design. Models having one child and having two children also illustrate the discontinuity of women's earnings due to having children at the 15% of significance level with their associated functions of tenure and treatment. As was expected in this study, treatment variables associated with having children demonstrate a significantly large earnings penalty for women (see Table 3.18).

Women aged between 40 and 55 Table 3.19 and Table 3.20 list the earnings effects associated with having children for women aged 40 and 55 around the cutoff points: 4 years in Korea and 12 years in Germany. The average tenures for this age group are 4.32 years in Korea and 12.40 years in Germany. In Korea, having 1 to 2 children results in a 37% positive earnings effect for women aged between 40 and 55 in the tenure bandwidth between 3 and 5 years up to 5% of significance level using the linear control function of tenure and treatment. Their control function also demonstrates highly significant effects with a value of 0.93 for tenure and a value of -0.47 for the interaction term of tenure and treatment. This result clearly illustrates the discontinuity of women's earnings due to having 1 to 2 children compared to having no children in Korea. A trend is thereby evident, having one child and having two children also support this discontinuity with 43% and 79% of the average treatment effects including the squared term. However, any positive effects that having children might have on female earnings are unexpected results in this study (see Table 3.19).

In Germany, having one child results in a -31% significant treatment effect on earnings for the model using the linear control function of tenure and treatment compared with women having no children. Having 1 to 2 children leads a -36% negative earnings effect using the linear control function and a -36% negative earnings effect including the squared term, when analyzing women who have worked for 11 to 13 years at their present job. These are much smaller than the effects of having children for women aged 30 and 40. The control functions, however, do not always demonstrate significant tenure earnings curves with/without treatment effect (see Table 3.20).

3.6. Conclusion

This study attempted to test the hypothesis that a discontinuity in a woman's career, measured by their earnings, could be attributable to the necessity of providing childcare in Korea and Germany, according to Mincer and Polachek (1974). Moreover, this study aimed to compare human capital and the earnings of women in both countries. The estimation results are arrived at by utilizing the quantile regression and the regression discontinuity design techniques. A summary of these follows below:

Having Children and Earnings Discontinuity Earnings effects due to children, for the most part, do not lead to significant results in Korea according to quantile regression estimation. Focusing on the significant results, women who have one child earn 5.4% more (25th quantile), women having two children earn -6.5% to -8% less, women with three children earn -10% to -17% less and those having four children earn 31% to 62% more compared to women having no children. The insignificant effects of children on earnings and the positive effects associated with having children on earnings in Korea are not typical. The home time in the human capital theory for women could not easily explain such phenomenon. In Germany, having children results in highly significant negative effects: -6% to -14% for women with one child, -11% to -24% for women with two children, -18% to -23% for women with three children, -12% to -23% for women with four children and -25% to -55.4% for women having five children. Women's home time necessitated by childcare certainly shows us an earnings penalty in Germany. The following trend can be noted: women at the lower quantiles suffer more from earnings loss due to children than women at the higher quantiles in Germany. They might have more difficulties in finding affordable childcare. For example, leaving their children in a day care center might be more difficult for women in the lower quantiles.

Using the regression discontinuity design, which aims to enhance the estimation results by removing the self-selection bias, this study found earnings discontinuities concerning children in some cases: Analyzing German women aged between 30 and 40, this study found a strong earnings loss (-65%) associated with having one to two children compared with women in the control group, who have no children. With noticeable negative effects, this verified suppositions about earnings discontinuity concerning children. This trend is also confirmed for young mothers of one or two children (about -49% or -70%). In Korea, women aged 40 and 55 who have one to two children earn 37% more than women who have no children, demonstrating significant earnings discontinuities due to children. By trend, women having one child or two children also experience earnings gains due to children: 43% or 79% respectively.

The negative effects of having children in Germany can be explained in the following way: Having children requires women to spend significant amounts of time at home for childbirth and child rearing. This home time, which disturbs their accumulated labor market experience and the acquisition of skills on the job, negatively affects their working careers in Germany. However, the positive earnings effects due to children in Korea could not be associated with theories of home time for women. In my opinion, there may be two explanations: First, the high educational costs and the strong social pressure for human capital investment in Korea financially burden the family having children. Especially, families which have children in junior/senior high school suffer from many expenses due to private tutoring fees for children.[26] Mothers aged 40 and 55, who may have children in junior/senior high school, want to supplement the high educational costs and might work harder than women without children. Second, mothers have a high reservation wage: They may consider the education of their children as the principal goal of their family. Thus, they are not willing to take a job in the labor market, if their husbands alone earn enough money to care for the family. Mothers aged between 40 and 55 could be especially concerned about the education of their children, who may be preparing to take an entrance exam for a university at this time.

26 Expenditure on educational institutions as a percentage of GDP for all levels of education in Korea amounts to 7.6% in 2003, which is the third highest across OECD-countries. Especially, the expenditure for private institutions in Korea is high. This reached 2.9% of GDP, which is the highest level across OECD-countries, while the expenditure for public institutions remains 4.6% of GDP, which just attains the 17th highest position in all OECD countries (OECD (2006)).

Family Responsibility and Effort on the Job Becker (1985) suggested that the specialization of house work and the low effort and attachment to the labor market of married women (with children) could be a reason for their low earnings. Using variables for marital status and family responsibility – such as female household head and single mother –, this study attempted to identify whether women who carry the burden of family responsibility make more effort on the job, which could enhance their earnings. Women who have family responsibility as household heads earn 7% to 9% more in Korea and 2.7% to 4% more in Germany than other women. Single mothers earn about 9% to 22% more in Korea and 7% to 24% more in Germany than other women. The burdens associated with family responsibility might force women to make more effort on the job in Korea and in Germany. This might lead to an increase in earnings for them.

Labor Market Experience In Korea, age in general does not have significant effects on women's earnings, while age has high significantly effects on earnings for German women. The dimension of earnings effects associated with age is also greater in Germany than in Korea. Tenure generally leads to significant earnings effects in both countries. Tenure-earnings profiles indicate a stronger effect on earnings in Korea than in Germany. However, the average tenure of Korean women amounts to only about 4 years and their median value to 2.3 years. Thus, their labor market experience along with age and tenure in general could not have a noticeable effect on the earnings of women in Korea.

Educational Earnings Differential The earnings differences associated with particular educational levels seem to be much larger in Germany than in Korea, compared to the base variable – senior high school graduation in Korea and a vocational qualification and/or qualification for college or university (*Berufsabschluss* and/or (*Fach-*)*Hochschulreife*) in Germany. This large earnings difference between educational levels in Germany might reflect individual and cultural divergences concerning women's education in German society such as the high proportion of immigrants in the population. The gain in earnings connected to a university level education – higher than a bachelor's degree in Korea and higher than a *Fachhochschuldiplom* in Germany – is higher in Korea than in Germany compared to each base variable. The effects of years of education show a similar trend in earnings as was observed with educational levels. 16 years of education (bachelor's degree in Korea) and 18 years+ of education (master's degree and higher in Korea) have a stronger effect on earnings in Korea than in Germany, when compared to 12 years of education. This could be caused by the relative productivity gap between university and senior high school level educations. And/or women with a higher educational level in Korea have higher reservation wages: Usually, women with a higher education level have higher family income due to their husbands, who are also highly educated and earn more than lower educated men. These mothers might decide to spend more time with their family and children.

Concluding this study, mothers aged between 30 and 40 in Germany and mothers aged between 40 and 55 in Korea have earnings/career discontinuities due to childcare according to the central hypothesis of this study. Using the regression discontinuity design technique, this study could demonstrate the breaks of tenure-earnings functions between women with and without children at 7 years of tenure in Germany and at 4 years of tenure in Korea. German mothers aged between 30 and 40 who have one or two children suffer from a strong earnings loss (-65%) compared to women who have no children. The working career of mothers, especially young mothers, is considerably disturbed by childbirth and rearing children in Germany, which might well explain why (qualified) women are afraid of having children. In Korea, mothers of one or two children between the ages of 40 and 55 enjoy higher earnings compared to women without children. This result does not seem to be plausible. A more unusual fact could be that: Although a university degree results in a higher earnings effect in Korea than in Germany, the participation rate of female university graduates in the labor market is lower than the German counterpart (OECD (2009a)). Labor market experience of women attributable to age is not significant in Korea. Their tenure has a much shorter duration in Korea than in Germany.

What are the reasons? Socially, Korean women seldom have a chance to continue their working career after childbirth. They could rarely utilize public services and any family friendly employment policies of their companies to balance their job and family life. Most women have to stay at home with the family after childbirth, which explains the short tenure of women in Korea. In this sense, women's careers are socially neglected and sacrificed for their family and children in Korea. However, Korean women do not acknowledge this problem as the most serious one in their family life. Koreans still tend to have a conviction that the common success of family takes precedence over the individual life of a family member. Parents are highly concerned for the education of their children due to significant social pressures concerning education in Korea. This may lead to a high reservation wage for women in Korea, which increases the earnings of mothers. In this regard, any future studies concerning the earnings of women need to consider the reservation wages related to children and family in Korea, which is not always possible in a comparative study, where the availability of variables is more limited. A study concerning the determinants of employment for more highly educated women in Korea would be especially interesting.

References

Beblo, M., Bender, S. and Wolf, E. (2006), "The Wage Effects of Entering Motherhood – A Within-Firm Matching Approach," ZEW Discussion Paper, 06-053, Mannheim.

Beblo, M. and Wolf, E. (2000), "How Much Does a Year off Cost? Estimating the Wage Effects of Employment Breaks and Part-Time Spells," ZEW Discussion Paper, 00-69, Mannheim.

Beblo, M. and Wolf, E. (2002), "The Wage Penalties for Heterogeneous Employment Biographies: An Empirical Analysis for Germany," ZEW Discussion Paper, 02-45, Mannheim.

Becker, G. (1985), "Human Capital, Effort and Sexual Division of Labor," *Journal of Labor Economics*, 3(1), pp. 33-58.

Ejrnaes, M. and Kunze, A. (2004), "Wage Dips and Drops around First Birth," Working Paper, 2004-01, Centre for Applied Microeconometrics, University of Copenhagen.

Galler, H.P. (1991), "Opportunitätskosten der Entscheidung für Familie und Haushalt," in: Gräbe, S. (Eds.), *Der private Haushalt als Wirtschaftsfaktor*, Frankfurt am Main: Campus, pp. 118-152.

Garen, J. (1984), "The Returns to Schooling: A Selectivity Bias Approach with a Continuous Choice Variable," *Econometrica*, 52(5), pp. 1199-1218.

Heckman, J.J. (1979), "Sample Selection Bias as a Specification Error," *Econometrica*, 47(1), pp. 153-161.

Heckman, J.J. and Robb, R. (1985), "Alternative Methods for Evaluating the Impact of Interventions," in: Heckman, J.J. and Singer, B. (Eds.), *Longitudinal Analysis of Labor Market Data*, Econometric Society Monographs Series, 10, Cambridge: Cambridge University Press, pp. 156-245.

Huh, S. and Hong, J.I. (2003), "Comparison of Gender Wage Differentials between Industries and Occupations," *Economic Research*, 18, Seoul: Chung-Ang University Economic Research Institute, pp. 111-132 (in Korean).

Kang, M.J. (2004), "Wage Effect of Women Reducing Working Time," *Women Studies Forum*, 5, Seoul: Korean Woman's Development Institute, pp. 30-34 (in Korean).

Kang, S.B. (2008), "Increase of Women's Employment in Types of Work and Wage Gaps," *Monthly Labor Review*, 42, Seoul: Korea Labor Institute, pp. 52-62 (in Korean).

Kim, J.H. (2009), "The Effects of Female Wage on Fertility in Korea," *KDI Journal of Economic Policy,* 2009(1), Seoul: Korea Development Institute, pp. 105-138 (in Korean).

Kim, S.H. (2009), "The Effect of Female Worker's Working Conditions on Fertility," *Social Welfare Policy*, 36(2), Seoul: Korea Association of Social Welfare Policy, pp. 113-137 (in Korean).

Kim, T.H. (2001), *Wages for Non-Regular Female Employees and Policy of Equal Pay for Equal Work*, Seoul: Ministry of Gender Equality & Family (in Korean).

Kim, Y.S. (2007), *A Study on Gender Income Gap*, Seoul: Korea Development Institute (in Korean).

Klaauw, W. (2002), "Estimating the Effect of Financial Aid Offers on College Enrollment: A Regression-Discontinuity Approach," *International Economic Review*, 43(4), pp. 1249-1287.

Koenker, R. (2005), *Quantile Regression*, Cambridge: Cambridge University Press.

Koenker, R. and Bassett, G. (1978), "Regression Quantiles," *Econometrica*, 46 (1), pp. 33-50.

Koenker, R. and Hallock, K.F. (2001), "Quantile Regression," *Journal of Economic Perspectives*, 15(4), pp. 143-156.

Korean Statistical Information Service (2010), *2009 Social Indicators in Korea*, Seoul: Statistics Korea (in Korean).

Kunze, A. (2002), "The Timing of Careers and Human Capital Depreciation," IZA-Discussion Paper, 509, Bonn.

Licht, G. and Steiner, V. (1991), "Male-Female Wage Differentials, Labor Force Attachment, and Human Capital Accumulation in Germany," *Volkswirtschaftliche Diskussionsreihe des Instituts für Volkswirtschaftslehre*, 65, Ausburg: Universität Augsburg.

Licht, G. and Steiner, V. (1992), "Individuelle Einkommensdynamik und Humankapitaleffekte nach Erwerbsunterbrechungen," *Jahrbuch für Nationalökonomie und Statistik*, 209(3-4), pp. 241-265.

Light, A. and Ureta, M. (1995), "Early-Career Work Experience and Gender Wage Differentials," *Journal of Labor Economics*, 13(1), pp. 121-154.

Machado, J.A.F. and Mata, J. (2001), "Earnings Functions in Portugal 1982-1994: Evidence from Quantile Regressions," *Empirical Economics*, 26(1), pp. 114-134.

Mincer, J. and Ofek, H. (1982), "Interrupted Work Careers: Depreciation and Restoration of Human Capital," *Journal of Human Resources*, 17(1), pp. 3-24.

Mincer, J. and Polachek, S. (1974), "Family Investments in Human Capital: Earnings of Women," *Journal of Political Economy*, 82(2), pp. 76-108.

Mincer, J. and Polachek, S. (1978), "Women's Earnings Reexamined," *Journal of Human Resources*, 13(1), pp. 118-134.

Moon, Y.K. and Kang, M.J. (2003), *Employment Effect of Women Reducing Working Time*, Seoul: Korean Woman's Development Institute (in Korean).

OECD (1999), *Classifying Educational Programmes – Manual for ISCED-97 Implementation in OECD Countries*, Paris.

OECD (2006), *Education at a Glance 2006: OECD Indicators*, Paris.

OECD (2007), *Babies and Bosses - Reconciling Work and Family Life (Vol. 5): A Synthesis of Findings for OECD Countries*, Paris.

OECD (2009a), *Education at a Glance 2009: OECD Indicators*, Paris.

OECD (2009b), *Society at a Glance 2009: OECD Indicators*, Paris.

Ondrich, J., Spiess, C.K. and Yang, Q. (2001), "The Effect of Maternity Leave on Women's Pay in Germany 1984-1994," Working Paper, Berlin: German Institute for Economic Research.

Park, K.J. (2008), *A Study of the Effect of Family Friendly Employment Policy on Childbirth and Child Rearing*, Dissertation, Myongji University (in Korean).

Park, S.I. (1982), "A Decomposition of Male/Female Wage Differential in Korea," *KDI Journal of Economic Policy*, 1982(6), Seoul: Korea Development Institute, pp. 59-87 (in Korean).

Polachek, S. and Kim, M.K. (1994), "Panel Estimates of the Gender Earnings Gap," *Journal of Econometrics*, 61(1), pp. 23-42 (in Korean).

Rosenbaum, P. and Rubin, D. (1983), "The Central Role of the Propensity Score in Observational Studies for Causal Effects," *Biometrika*, 70, pp. 41-55.

Schönberg, U. and Ludsteck, J. (2007), "Maternity Leave Legislation, Female Labor Supply, and the Family Wage Gap," IZA Discussion Paper, 2699, Bonn.

Shin, K.S. (2003), *A Study on Occupational Sex Segregation and Gender Wage Gap in Korean Labor Market*, Dissertation, Dongguk University (in Korean).

Son, J.Y. (2009), "The Development of the Real Estate Market and the Policy Issues," *CFE Report*, 106, Seoul: Center for Free Enterprise (in Korea).

Thistlewaithe, D.L. and Campbell, D.T. (1960), "Regression-Discontinuity Analysis: An Alternative to the Ex Post Facto Experiment," *Journal of Educational Psychology*, LI, pp. 309-317.

Trochim, M.K. (1984), *Research Design for Program Evaluation: the Regression-Discontinuity Approach*, William Beverly Hills: Sage.

Yun, B.J. (1990), "Gender Wage Differentials," *Business Research*, 14, Seoul: Dongguk Graduates School, pp. 107-118 (in Korean).

Appendix

Table 3.1: Descriptive Statistics for the Korean Data KLIPS for Women (3,066 observations of women aged between 30 and 55)

Variable	Mean	Std. Dev	Min	Max
Ln earnings	9.172622	.6056827	5.132261	11.34776
Yearly gross earnings in Euro	11,544.6	7,848.941	169.3996	84,775.7
Age	41.05969	7.035127	30	55
Tenure	4.209257	4.959328	0	35.41667
Children				Percent
0				42.04
1				20.22
2				33.76
3				3.91
4				0.07
Marital				Percent
Single				8.22
Married				80.01
Widowed				6.03
Divorced				4.11
Separated				1.63
Household head				15.20
Single mother				3.98
Complete education level				Percent
Under Junior high school				12.33
Junior high school				19.60
Senior high school				45.60
Junior college				8.94
University +				13.54
Completed job training				7.70
Public sector				3.62
Overtime				21.33

Part time				10.89
Firm size				Percent
under 5 employees				22.64
5-9				15.13
10-99				31.57
100-999				15.72
1000 and more				14.94
Residence				Percent
Seoul				23.16
Busan				11.71
Daegu				6.16
Deajeon				2.87
Incheon				7.53
Gwangju				2.25
Ulsan				3.00
Gyunggi				21.46
Gangwon				1.24
Chungbuk				2.38
Chungnam				2.94
Jeonbuk				2.15
Jeonnam				2.61
Gyungbuk				4.40
Gyungnam				6.13
Jeju				0
Year				Percent
2002				17.55
2003				17.91
2004				20.32
2005				22.02
2006				22.21

Note *The description of 11 occupational dummies and 33 industry dummies is omitted.*

Table 3.2: Descriptive Statistics for the German Data GSOEP for Women (8,260 observations of women aged between 30 and 55)

Variable	**Mean**	**Std. Dev.**	**Min**	**Max**
Ln earnings	9.781201	.8280458	5.634789	12.00665
Yearly gross earnings in Euro	22,536.86	15,073.85	50	303,200
Age	42.12288	6.957365	30	55
Tenure	10.21734	8.790374	0	40.8
Children				Percent
0				52.68
1				24.08
2				18.80
3				3.72
4				0.67
5				0.06
Marital				Percent
Single				18.20
Married				66.04
Widowed				1.42
Divorced				11.76
Separated				2.59
Household head				43.33
Single mother				10.40
Complete education level				Percent
Under Real/Haupt				0.59
Real/Hauptschule				11.91
Berufsabschluss and/or (Fach-)Hochschulreife				61.92
Technik/ Meisterschule				8.78
Fachhochschul+				16.79
Completed job training				0.41
Public sector				35.67

Overtime				44.12
Part time				56.90
Firm size				Percent
under 5 employees				10.62
5-19				17.75
20-99				18.92
100-1999				31.13
2000 and more				21.59
Residence				Percent
Berlin (West)				2.36
Schleswig-Hols				3.81
Hamburg				1.73
Lower Saxony				11.55
Bremen				0.92
N-Rhein-Westfa				28.38
Hessen				8.83
R-Pfalz,Saarl.				8.50
Baden-Wuerttemb.				14.16
Bavaria				18.44
Berlin (East)				0.22
Mecklenburg-V.				0.22
Brandenburg				0.31
Saxony-Anhalt				0.11
Thueringen				0.12
Saxony				0.34
Year				Percent
2002				23.08
2003				21.20
2004				21.31
2005				16.83
2006				17.59

Note *The description of 11 occupational dummies and 30 industry dummies is omitted.*

Table 3.3: Descriptive Statistics of Main Variables after Treatment Assignment for Each Model in Korea

Korea		**All tenure**					**+/-1 year around cut-off point: Tenure 4 year**				
		Obs.	Ln earnings (mean)	Age (mean year)	Tenure (mean year)	Treatment=1 (%)	Obs.	Ln earnings (mean)	Age (mean year)	Tenure (mean year)	Treatment=1 (%)
Aged 30-55	**1 Child**	917	9.23466	42.16467	6.370138	42.20	152	9.221344	42.74342	4.078947	40.13
	2 Child	1311	9.136378	40.36308	4.823036	59.57	186	9.168608	41.26344	3.954749	51.08
	1-2 Child	1698	9.125098	40.27326	4.073812	68.79	247	9.187675	41.24696	3.840756	63.16
Aged 30-40	**1 Child**	351	9.411246	33.72365	5.242403	52.14	48	9.581405	33.39583	4.050347	41.67
	2 Child	702	9.134859	35.04558	3.26745	76.07	90	9.229847	35.23333	3.742593	68.89
	1-2 Child	885	9.146148	34.96045	2.878249	81.02	110	9.298653	35.11818	3.706818	74.55
Aged 40-55	**1 Child**	590	9.128505	47.09831	6.912429	37.80	108	9.066071	46.7963	4.094136	39.81
	2 Child	674	9.135205	45.86647	6.170129	45.55	107	9.105983	46.20561	4.101246	39.25
	1-2 Child	897	9.10186	45.48941	5.05379	59.09	150	9.102706	45.63333	3.921667	56.67

Note *Treatment equals one, if women have children.*

Table 3.4: Descriptive Statistics of Main Variables after Treatment Assignment for Each Model in Germany

Germany		All tenure					+/-1 year around cutoff point of tenure				
		Obs.	Ln earnings (mean)	Age (mean year)	Tenure (mean year)	Treatment=1 (%)	Obs.	Ln earnings (mean)	Age (mean year)	Tenure (mean year)	Treatment=1 (%)
Aged 30-55 cutoff point: 10 years	**1 Child**	3340	9.879793	44.13263	13.93132	36.86	225	10.00173	43.39111	10.17111	38.67
	2 Child	3194	9.822721	43.81841	14.21775	33.97	184	9.92993	43.6413	10.31141	25.00
	1-2 Child	4425	9.726068	42.86599	11.39055	52.34	271	9.914003	42.88561	10.05277	49.08
Aged 30-40 cutoff point: 7 years	**1 Child**	1179	9.872213	35.40628	8.536302	44.53	116	10.04162	35.19828	7.167241	36.21
	2 Child	1276	9.695597	35.69201	7.989812	48.75	117	10.02063	35.50427	7.204274	36.75
	1-2 Child	1801	9.604653	35.66185	6.509717	63.69	159	9.885923	35.77358	7.001258	53.46
Aged 40-55 cutoff point: 12 years	**1 Child**	2262	9.893547	47.56454	16.42767	33.82	155	9.961365	47.6129	12.28258	27.10
	2 Child	2021	9.915558	47.46264	17.67091	25.93	128	9.982925	47.70313	12.45938	11.72
	1-2 Child	2786	9.804917	46.73475	14.20122	46.27	170	9.942209	47.24706	12.22471	33.53

Note *Treatment equals one, if women have children.*

Table 3.5: Estimation Results for Women – Age and Tenure in Korea

Korea	OLS	0.1	0.25	Median	0.75	0.9
Earnings						
Age	.018597 1.18	.043349 1.41	.036586 2.26	0.029861 1.37	0.005217 0.31	.0089064 0.55
Age squared	-.00026 -1.40	-.00058 -1.59	-.00048 -2.51	-.0004 -1.53	-.00013 -0.64	-.0002 -1.01
Tenure	.04349 9.80	.05921 6.99	.04117 9.14	.038835 6.44	.0317 7.22	.025919 5.96
Tenure squared	-.0005 -2.33	-.00136 -3.56	-.00045 -2.14	-.00041 -1.43	.000131 0.66	.000295 1.59
Obs.	3066					
(Psd.) R-sq	0.5078	0.2652	0.2782	0.3313	0.4040	0.4330

Note *t-statistics are presented in the 2nd. line of cells in the table.*

Table 3.6: Estimation Results for Women – Age and Tenure in Germany

Germany	OLS	0.1	0.25	Median	0.75	0.9
Earnings						
Age	.096948 7.67	.096419 3.95	.071383 5.08	.078009 8.49	.068118 5.80	.070204 6.28
Age squared	-.00106 -7.43	-.00107 -3.88	-.00077 -4.80	-.00084 -7.98	-.00075 -5.52	-.00078 -6.06
Tenure	.027813 11.81	.052607 12.10	.031611 12.37	.019409 11.26	.014065 6.26	.010426 4.74
Tenure squared	-.00053 -7.05	-.00111 -8.53	-.00066 -8.31	-.00040 -7.28	-.00022 -2.97	-.0001 -1.40
Obs.	8260					
(Psd.) R-sq	0.5477	0.3877	0.3978	0.3628	0.3298	0.3039

Note *t-statistics are presented in the 2nd. line of cells in the table*

Figure 3.5: Age-Earnings Profiles for Women in Korea

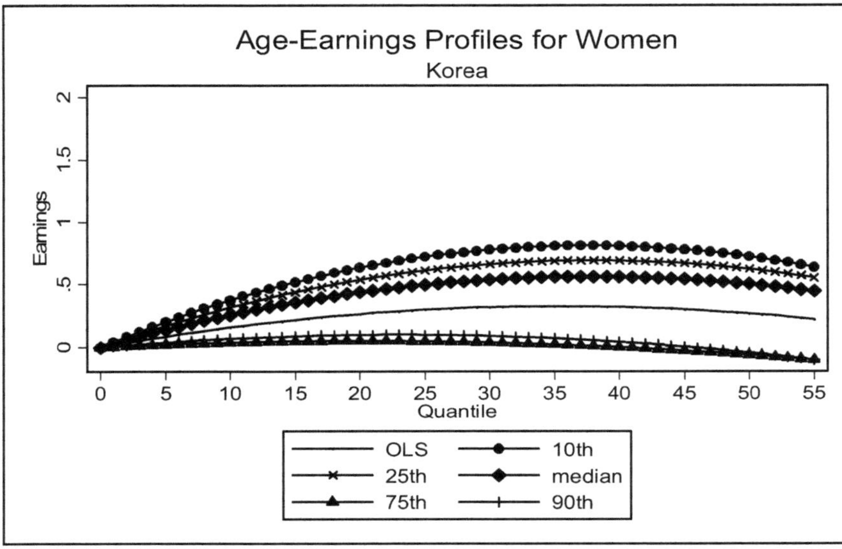

Figure 3.6: Age-Earnings Profiles for Women in Germany

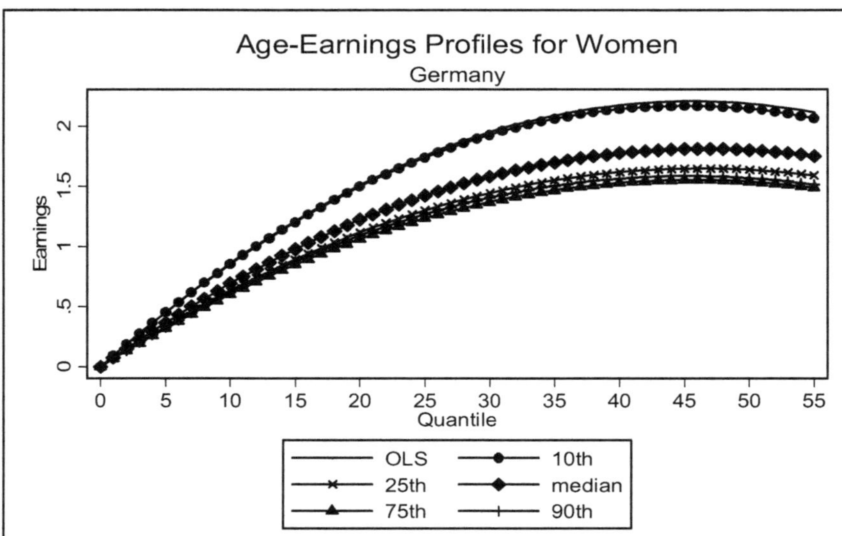

Figure 3.7: Tenure-Earnings Profiles for Women in Korea

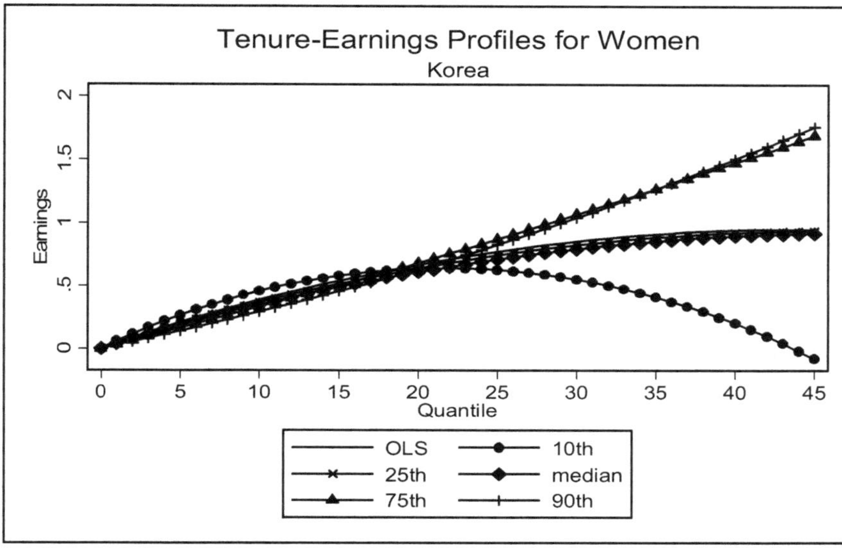

Figure 3.8: Tenure-Earnings Profiles for Women in Germany

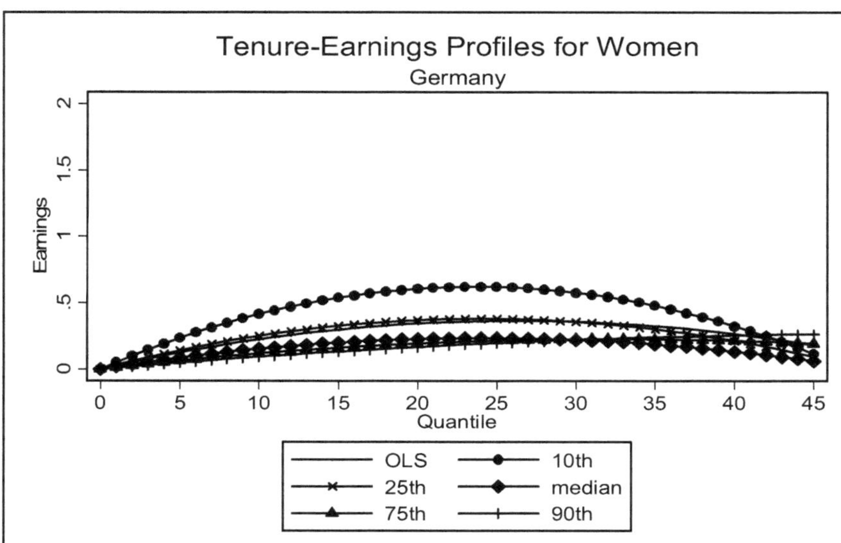

Table 3.7: Estimation Results for Women – Number of Children in Korea

Korea	OLS	0.1	0.25	Median	0.75	0.9
Earnings						
1 Child	.011244 0.41	.010231 0.20	.053605 1.92	.009888 0.26	.027996 0.96	-.00775 -0.28
2 Children	-.031126 -1.19	-.08202 -1.76	-.02444 -0.92	.002635 0.07	-.01893 -0.66	-.06526 -2.41
3 Children	-.07721 -1.74	-.09913 -1.18	-.09663 -2.20	-.05788 -0.93	-.04549 -0.97	-.1674 -4.03
4 Children	.446454 1.19	.623412 2.64	.477874 2.81	.177114 0.53	.306392 1.82	-.12659 -1.38
5 Children	(dropped)	(dropped)	(dropped)	(dropped)	(dropped)	(dropped)
Obs.	3066					
(Psd.)R-sq	0.5078	0.2652	0.2782	0.3313	0.4040	0.4330

Note *Base variable for number of children dummies: Having no children. t-statistics are presented in the 2nd. line of cells in the table.*

Table 3.8: Estimation Results for Women – Number of Children in Germany

Germany	OLS	0.1	0.25	Median	0.75	0.9
Earnings						
1 Child	-.10894 -5.72	-.13244 -3.84	-0.14076 -6.88	-.10757 -7.70	-.09704 -5.17	-0.05799 -3.12
2 Children	-.19028 -9.75	-.18271 -5.14	-.23606 -11.42	-.17494 -12.10	-.16677 -8.61	-.10653 -5.39
3 Children	-.2145 -6.64	-.18977 -3.11	-0.22903 -6.53	-.17923 -7.39	-.1914 -6.12	-.18416 -6.08
4 Children	-.19009 -2.69	-.23265 -1.85	-.23076 -3.07	-.12289 -2.33	-.04813 -0.67	-.01143 -0.18
5 Children	-.41785 -2.14	-.44378 -3.74	-.55421 -3.19	-.2513 -1.73	-.30493 -1.67	-.39738 -6.66
Obs.	8260					
(Psd.)R-sq	0.5477	0.3877	0.3978	0.3628	0.3298	0.3039

Note *Base variable for number of children dummies: Having no children. t-statistics are presented in the 2nd. line of cells in the table.*

Figure 3.9: Effects of Having Children for Women in Korea

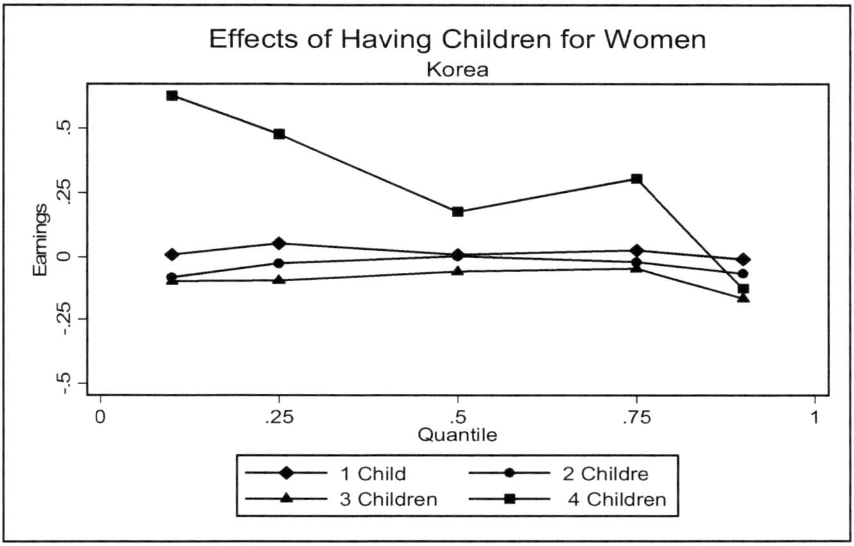

Figure 3.10: Effects of Having Children for Women in Germany

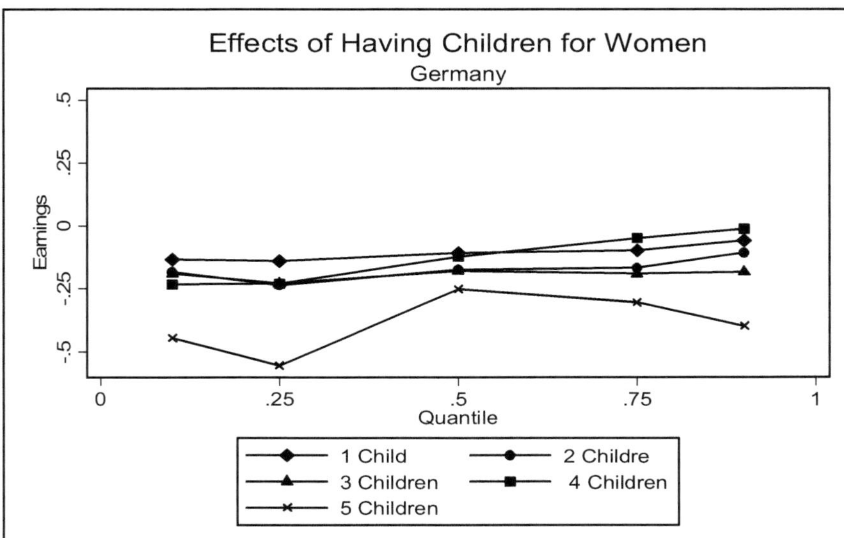

Table 3.9: Estimation Results for Women – Family Characteristics in Korea

Korea	OLS	0.1	0.25	Median	0.75	0.9
			Earnings			
Married	.046573 1.17	.018508 0.25	.03019 0.76	.003796 0.07	.025196 0.61	.107574 2.53
Widowed	-.0623 -1.15	-.14821 -1.53	-.03128 -0.57	-.06814 -0.92	-.00815 -0.14	.011923 0.19
Divorced	-.015108 -0.25	-.00063 -0.01	-.03417 -0.59	-.05907 -0.73	.080108 1.20	.063519 0.93
Separated	.14588 1.77	-.03962 -0.28	.047899 0.59	.122244 1.10	.204565 2.51	.336432 3.82
Household head	.088055 2.39	.089866 1.21	.088975 2.39	.070461 1.40	.021068 0.58	.038093 1.06
Single mother	.098247 1.72	.217166 1.88	.087655 1.55	.078976 1.03	.035483 0.64	.101456 1.80
Obs.				3066		
(Psd.)R-sq	0.5078	0.2652	0.2782	0.3313	0.4040	0.4330

Note *Base variable for marital status: Single. t-statistics are presented in the 2nd. line of cells in the table.*

Figure 3.11: Effects of Marriage for Women

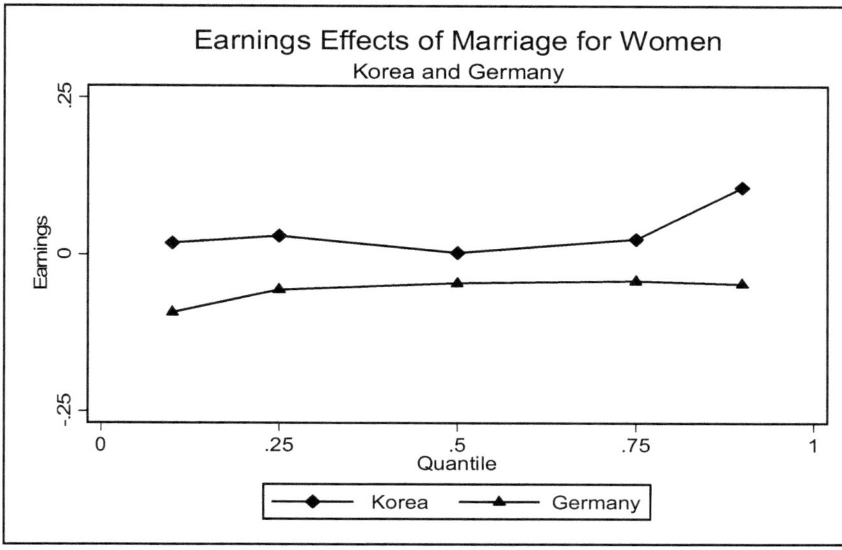

Table 3.10: Estimation Results for Women – Family Characteristics in Germany

Germany	OLS	0.1	0.25	Median	0.75	0.9
			Earnings			
Married	-.06302 -2.89	-.09318 -2.32	-.05689 -2.36	-.04554 -2.82	-.04132 -1.94	-.04616 -2.30
Widowed	-.04136 -0.76	-.06267 -0.62	-.03027 -0.49	-.0794 -2.03	-.0842 -1.65	-.07978 -1.65
Divorced	.0233 0.88	.068908 1.37	.022502 0.76	-.00309 -0.16	.004431 0.18	.019048 0.81
Separated	-.02846 -0.67	-.03121 -0.41	-.06895 -1.51	-.0311 -1.01	-.02221 -0.56	.012473 0.34
Household head	.02677 1.77	-.0159 -0.57	.029045 1.72	.007659 0.70	.017033 1.19	.037907 2.76
Single mother	.171348 5.32	.236259 3.82	.219261 5.98	.179091 7.57	.166925 5.51	.072454 2.62
Obs.				8260		
(Psd.)R-sq	0.5477	0.3877	0.3978	0.3628	0.3298	0.3039

Note *Base variable for marital status: Single, t-statistics are presented in the 2nd. line of cells in the table.*

Figure 3.12: Effects of Female Household Head

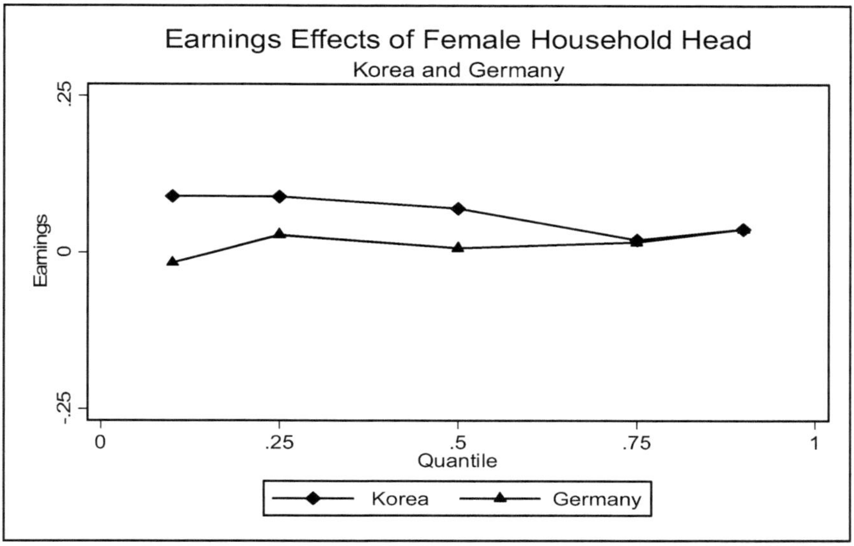

Figure 3.13: Effects of Single Mother

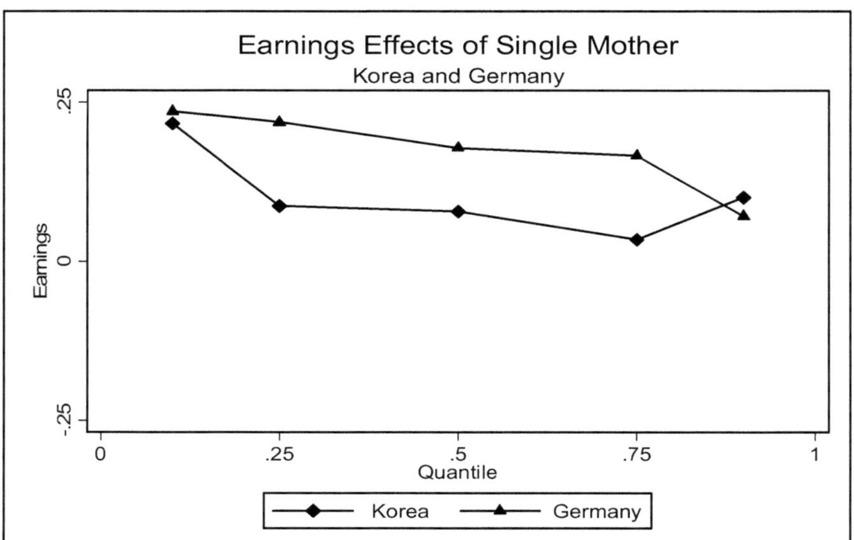

Table 3.11: Estimation Results for Women – Educational Levels in Korea

Korea	OLS	0.1	0.25	Median	0.75	0.9
			Earnings			
Under junior high	-.09526 -3.27	-.06807 -1.16	-.08384 -2.78	-.10364 -2.61	-.08569 -2.81	-.1074 -3.38
Junior high school	.028074 1.13	.007129 0.14	.008873 0.35	-.008441 -0.25	.008104 0.32	.012594 0.49
Junior College	.078509 2.24	.106506 1.54	.111548 3.07	.110867 2.27	.108149 2.88	.09457 2.37
University+	.397678 10.63	.347667 4.50	.360813 9.49	.442575 8.42	.472829 12.18	.452899 11.67
Obs.			3066			
(Psd.)R-sq	0.5078	0.2652	0.2782	0.3313	0.4040	0.4330

Note *Base variable for education level: Senior high school. t-statistics are presented in the 2nd. line of cells in the table.*

Table 3.12: Estimation Results for Women – Educational Levels in Germany

Germany	OLS	0.1	0.25	Median	0.75	0.9
			Earnings			
Under Real/Hauptschule	-.65343 -12.72	-.83745 -12.88	-.78073 -17.01	-.54845 -13.08	-.45943 -7.68	-.40988 -7.31
Real/Hauptschule	-.08355 -4.24	-.13097 -3.71	-.07111 -3.22	-.0479 -3.25	-.03938 -2.06	-.06362 -3.54
Technik/ Meisterschule	.077599 3.16	.126766 2.59	.168911 5.91	.109345 5.98	.04621 2.00	.025259 1.17
Fachhochschul+	.248082 9.76	.267731 4.91	.280517 9.59	.223428 12.11	.258905 10.68	.261123 11.17
Obs.			8260			
(Psd.)R-sq	0.5477	0.3877	0.3978	0.3628	0.3298	0.3039

Note *Base variable for education level: Berufsabschluss and/or (Fach-)Hochschulreife. t-statistics are presented in the 2nd. line of cells in the table.*

Table 3.13: Estimation Results for Women – Educational Years in Korea

Korea	OLS	0.1	0.25	Median	0.75	0.9
Earnings						
0 years	-.16568 -2.35	.015691 0.12	-.18422 -2.74	-.21647 -3.49	-.18483 -2.19	-.21659 -2.55
6 years	-0.08773 -2.93	-0.06938 -1.32	-0.08446 -2.87	-0.09669 -3.51	-0.07205 -2.03	-0.07251 -1.84
9 years	.028285 1.14	.026975 0.58	.009888 0.41	-.00428 -0.19	.000107 0.00	0.037107 1.15
14 years	.083246 2.37	.124274 2.00	.11487 3.25	.127275 3.84	.099719 2.36	.097437 2.00
16 years	.383327 10.05	.350174 5.00	.352511 9.35	.418561 11.68	.420025 9.33	.417815 8.99
18 + years	.517018 6.55	.193245 1.58	.565356 7.38	.667283 8.70	.623903 6.32	.521188 5.35
Obs.	3065					
(Psd.) R-sq	0.5084	0.2659	0.2788	0.3333	0.4055	0.4336

Note *Base variable for educational years: 12 years education, t-statistics are presented in the 2nd. line of cells in the table.*

Figure 3.14: Effects of Educational Years for Women in Korea

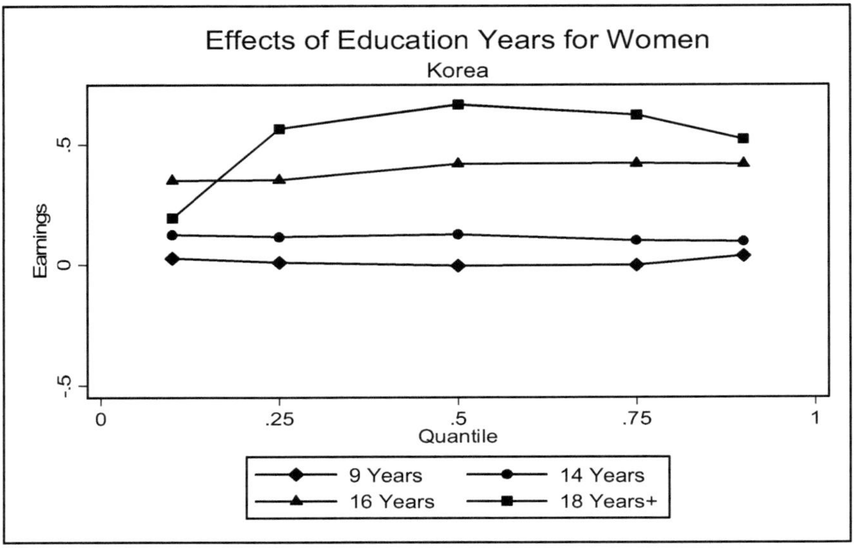

Figure 3.15: Effects of Educational Years for Women in Germany

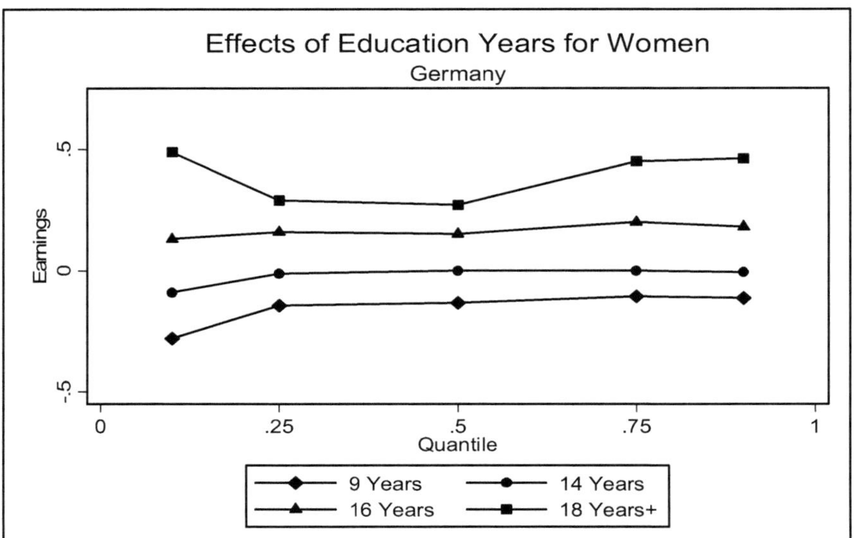

Table 3.14: Estimation Results for Women – Educational Years in Germany

Germany	OLS	0.1	0.25	Median	0.75	0.9
Earnings						
7 years	-.68592 -13.10	-.87225 -12.58	-.83458 -16.41	-.59452 -13.30	-.48835 -9.51	-.48827 -6.82
8.5 years	-.28934 -3.05	-.75149 -6.97	-.19844 -1.60	-.02263 -0.27	-.05558 -0.66	-.03409 -0.30
9 years	-.13685 -4.65	-.28026 -5.17	-.14336 -3.78	-.13276 -5.83	-.10624 -4.51	-.11451 -3.34
10 years	-.09565 -2.51	-.17543 -2.47	-.04875 -0.98	-.05132 -1.70	.005265 0.17	-.00334 -0.07
10.5 years	-.06617 -2.86	-.09017 -1.96	-.06029 -2.01	-.1006 -5.74	-0.07663 -4.23	-0.0466 -1.77
11 years	-.09794 -3.07	-.15497 -2.55	-.13116 -3.28	-.08581 -3.45	-.04798 -1.85	-.01689 -0.45
11.5 years	-.001401 -0.06	-.044352 -0.99	-.05022 -1.77	-.02929 -1.70	.035028 1.93	.03386 1.31
13 years	-.13064 -2.57	-.25119 -2.66	-.13325 -2.05	-.10378 -2.60	.029868 0.69	.039907 0.63
13.5 years	.016712 0.34	-.0349 -0.37	-.02795 -0.46	.010097 0.27	.105184 2.63	.139639 2.52
14 years	-.04703 -0.92	-.09029 -0.93	-.01076 -0.16	-.00054 -0.01	.000402 0.01	-.00667 -0.12
14.5 years	.094584 2.53	.071564 0.97	.072834 1.54	.115191 3.94	.157154 5.16	.119227 2.87
15 years	.103523 2.84	.071171 1.00	.072787 1.58	.119333 4.19	.163788 5.60	.173923 4.10
16 years	.162564 3.89	.131529 1.62	.159731 3.01	.151725 4.72	.200571 5.98	.181253 3.90
17 years	.163417 1.55	.179427 0.90	.237229 1.79	.181618 2.25	.111424 1.41	.051666 0.48
18+ years	.348714 9.12	.4872 5.74	.289838 6.01	.27241 9.50	.451509 14.10	.461213 9.63
Obs.	8217					
(Psd.)R-sq	0.5533	0.3953	0.4003	0.3674	0.3381	0.3116

Note *Base variable for educational years: 12 years education. t-statistics are presented in the 2nd. line of cells in the table.*

Table 3.15: Estimation Results – Having Children in Korea (Women between the Ages of 30 and 55, Cutoff Point: 4 Years)

Korea	All tenure						Tenure: 3-5					
	1 Child		2 Children		1-2 Children		1 Child		2 Children		1-2 Children	
	Linear	Square	Linear	Square	Linear	Square	Linear	Square	Linear	Square	Linear	Square
	Earnings											
Treatment	.224231 3.23	.141297 1.50	-.03847 -0.67	-.03633 -0.46	.070642 1.34	.045567 0.65	.135811 0.99	.225487 0.96	-.0773 -0.50	.049274 0.17	.149079 1.08	.284971 1.09
Tenure	.027045 6.87	.033168 3.47	.0252334 6.47	.034271 3.55	.026491 6.64	.035994 3.66	.282847 1.81	4.194218 2.07	.214305 1.15	.907622 0.68	.346607 1.85	2.40525 1.48
Tenure Squared		-.00031 -0.69		-.00044 -0.97		-.00048 -1.03		-.72282 -1.78		-.33461 -0.49		-.57185 -1.15
Tenure·treatment	.09873 4.82	-.02661 -0.33	.012343 0.82	-.02228 -0.36	.040466 3.18	-.0306 -0.61	-.25911 -1.41	-.96508 -3.36	-.20338 -1.05	-.40288 -0.46	-.25479 -1.51	-0.8414 -1.99
Tenuresq·treatment		-.02574 -1.46		-.00537 -0.41		-.01324 -1.21		-.37221 -0.56		.762452 0.55		.283805 0.29
Obs.	917	917	1311	1311	1698	1698	152	152	186	186	247	247
R-sq	0.5228	0.5243	0.5004	0.5009	0.4469	0.4478	0.7943	0.8240	0.6925	0.6934	0.6348	0.6424

Note *Treatment equals one if women have 1, 2 or 1-2 children. Treatment equals zero, if women have no children. Tenure·treatment is the interaction term of tenure and treatment, and Tenuresq·treatment is the interaction term of tenure squared and treatment. t-statistics. These are presented in the 2nd. line of cells in the tables.*

Table 3.16: Estimation Results – Having Children in Germany (Women between the Ages of 30 and 55, Cutoff Point: 10 Years)

Germany	All tenure						Tenure: 9-11					
	1 Child		2 Children		1-2 Children		1 Child		2 Children		1-2 Children	
	Linear	Square	Linear	Square	Linear	Square	Linear	Square	Linear	Square	Linear	Square
	Earnings											
Treatment	-.03848 -0.90	-.08266 -1.38	-.12477 -2.63	-.22266 -3.24	-.06424 -1.68	-.12582 -2.36	-.01829 -0.12	-.06840 -0.29	-.05769 -0.27	-.1288 -0.40	-.04328 -0.30	-.15403 -0.73
Tenure	.004204 2.44	-.01342 -2.39	.003479 1.99	-.01008 -1.77	.002698 1.49	-.01182 -1.97	.07123 0.42	-.04216 -0.05	.126108 0.65	.078878 0.08	.070237 0.40	-.0311 -0.04
Tenure Squared		.000734 3.28		.000565 2.49		.000604 2.54		.096146 0.14		.032416 0.04		.080302 0.11
Tenure· treatment	.040557 6.96	.072012 3.05	.046951 7.00	.029652 1.07	.045409 9.37	.055017 2.80	.078 0.33	-.03146 -0.03	.302083 0.80	-.06204 -0.05	.203165 0.85	-.35968 -0.47
Tenuresq· treatment		.000442 0.22		-.00329 -1.39		-.0011 -0.66		-.32257 -0.36		-.32455 -0.31		-.56084 -0.98
Obs.	3340	3340	3194	3194	4425	4425	225	225	184	184	271	271
R-sq	0.5965	0.5978	0.6246	0.6255	0.5909	0.5915	0.6818	0.6821	0.7210	0.7213	0.6718	0.6755

Note *Treatment equals one if women have 1, 2 or 1-2 children. Treatment equals zero, if women have no children. Tenure·treatment is the interaction term of tenure and treatment, and Tenuresq·treatment is the interaction term of tenure squared and treatment. t-statistics. These are presented in the 2nd. line of cells in the tables.*

Table 3.17: Estimation Results – Having Children in Korea (Women between the Ages of 30 and 40, Cutoff Point: 4 Years)

Korea	All tenure						Tenure: 3-5					
	1 Child		2 Children		1-2 Children		1 Child		2 Children		1-2 Children	
	Linear	Square	Linear	Square	Linear	Square	Linear	Square	Linear	Square	Linear	Square
	Earnings											
Treatment	.230126 1.41	.172899 0.84	-.2298 -2.19	-.21619 -1.72	-.13748 -1.28	-.12523 -0.97	(dropped)	(dropped)	-.16344 -0.54	-.40426 -0.73	-.0044 -0.01	.008505 0.01
Tenure	.028456 2.09	.026418 0.67	.01471 1.25	.026854 0.75	.015138 1.25	.022941 0.61	(dropped)	(dropped)	.147578 0.40	-.50421 -0.29	.144068 0.38	.614765 0.21
Tenure Squared		.000175 0.05		-.00111 -0.36		-.00071 -0.22		(dropped)		1.060757 0.34		-.26807 -0.15
Tenure·treatment	.157283 4.06	.095513 0.62	.009975 0.44	.004297 0.05	.048897 2.28	.04757 0.56	(dropped)	(dropped)	-.16338 -0.42	-.07216 -0.03	-.30983 -1.30	-.57559 -0.35
Tenuresq·treatment		-.0128 -0.41		.002539 0.14		.002136 0.13		(dropped)		-.75935 -0.58		-.08105 -0.04
Obs.	351	351	702	702	885	885	48	48	90	90	110	110
R-sq	0.5760	0.5762	0.5554	0.5555	0.4892	0.4893			0.8701	0.8728	0.7913	0.7925

Note *Treatment equals one if women have 1, 2 or 1-2 children. Treatment equals zero, if women have no children. Tenure·treatment is the interaction term of tenure and treatment, and Tenuresq·treatment is the interaction term of tenure squared and treatment. t-statistics. These are presented in the 2nd. line of cells in the tables.*

Table 3.18: Estimation Results – Having Children in Germany (Women between the Ages of 30 and 40, Cutoff Point: 7 Years)

Germany	All tenure						Tenure: 6-8					
	1 Child		2 Children		1-2 Children		1 Child		2 Children		1-2 Children	
	Linear	Square	Linear	Square	Linear	Square	Linear	Square	Linear	Square	Linear	Square
	Earnings											
Treatment	-.41412 -6.61	-.35671 -3.94	-.41349 -5.99	-.37537 -3.98	-.39089 -6.46	-.34424 -4.19	-.13931 -0.58	-.49339 -1.50	-.34071 -1.22	-.6974 -2.03	-.17273 -0.69	-.64565 -2.21
Tenure	-.01681 -3.01	.001783 0.10	-.01374 -2.32	.000439 0.02	-.0147 -2.44	.001871 0.10	.033021 0.14	-.95303 -2.14	.55670 1.37	-.90728 -1.22	.098111 0.31	-.96707 -2.06
Tenure Squared		-.00136 -1.16		-.00104 -0.83		-.0012 -0.93		15.08559 2.19		10.6889 1.47		22.67908 2.16
Tenure·treatment	.041931 3.05	.06131 1.11	.064981 4.68	.070771 1.33	.055446 4.87	.065461 1.48	.265915 0.58	55.5754 2.07	-.42086 -1.16	2.218174 0.43	-.0069 -0.02	21.75452 1.60
Tenuresq·treatment		.00639 0.95		.003971 0.61		.004703 0.91		-.87739 -1.31		-.9757 -1.46		-.97073 -2.10
Obs.	1179	1179	1276	1276	1801	1801	116	116	117	117	159	159
R-sq	0.6536	0.6542	0.6706	0.6709	0.6269	0.6272	0.9262	0.9346	0.8267	0.8364	0.7920	0.8041

Note *Treatment equals one if women have 1, 2 or 1-2 children. Treatment equals zero, if women have no children. Tenure·treatment is the interaction term of tenure and treatment, and Tenuresq·treatment is the interaction term of tenure squared and treatment. t-statistics. These are presented in the 2nd. line of cells in the tables.*

Table 3.19: *Estimation Results – Having Children in Korea (Women between the Ages of 40 and 55, Cutoff Point: 4 Years)*

Korea	All tenure						Tenure: 3-5					
	1 Child		2 Children		1-2 Children		1 Child		2 Children		1-2 Children	
	Linear	Square	Linear	Square	Linear	Square	Linear	Square	Linear	Square	Linear	Square
	Earnings											
Treatment	.228417 2.98	.159771 1.55	.085985 1.03	.038158 0.36	.173083 2.72	.109023 1.33	.219385 1.15	.435048 1.30	.212889 0.90	.786986 1.62	.371332 2.12	.422625 1.40
Tenure	.030662 7.45	.026693 2.42	.027295 6.17	.034298 2.96	.027321 6.61	.034234 3.05	.464082 1.88	6.62831 2.09	.948354 2.98	8.26623 1.92	.927384 3.48	2.95550 1.56
Tenure Squared		.000197 0.39		-.00033 -0.62		-.00033 -0.63		-.7832 -1.76		-0.7611 -1.37		-.4898 -0.86
Tenure· Treatment	.029451 1.24	-.03644 -0.40	.012099 0.55	-.07123 -0.86	.037951 2.30	-.06336 -1.01	-.39248 -1.86	-.96851 -2.62	-.37225 -1.51	-0.8167 -1.17	-.47141 -2.78	-.89301 -2.19
Tenuresq· treatment		-.01663 -0.79		-.01774 -0.94		-.02172 -1.52		.201161 0.17		4.60290 1.15		-.17599 -0.21
Obs.	590	590	674	674	897	897	108	108	107	107	150	150
R-sq	0.5560	0.5567	0.5266	0.5276	0.4674	0.4692	0.8101	0.8343	0.8537	0.8614	0.7697	0.7804

Note *Treatment equals one if women have 1, 2 or 1-2 children. Treatment equals zero, if women have no children. Tenure·treatment is the interaction term of tenure and treatment, and Tenuresq·treatment is the interaction term of tenure squared and treatment. t-statistics. These are presented in the 2nd. line of cells in the tables.*

Table 3.20: Estimation Results – Having Children in Germany (Women between the Ages of 40 and 55, Cutoff Point: 12 Years)

Germany	All tenure						Tenure: 11-13					
	1 Child		2 Children		1-2 Children		1 Child		2 Children		1-2 Children	
	Linear	Square	Linear	Square	Linear	Square	Linear	Square	Linear	Square	Linear	Square
	Earnings											
Treatment	-.0038 -0.07	-.09283 -1.26	-.16351 -2.74	-.35106 -4.37	-.03661 -0.78	-.16392 -2.36	-.31253 -1.97	-.36535 -1.42	-.29153 -1.09	-.11555 -0.22	-.35663 -2.40	-.3637 -1.48
Tenure	.005961 2.83	-.00893 -1.29	.00604 2.95	-.00635 -0.95	.005107 2.37	-.00607 -0.85	-.17578 -1.15	-.10803 -0.15	-.22851 -1.32	.027116 0.03	-.20968 -1.32	-.32515 -0.47
Tenure Squared		.000652 2.25		.000547 1.96		.000489 1.63		-.06559 -0.10		-.22486 -0.30		.149363 0.19
Tenure·treatment	.027772 4.50	.025722 1.02	.020213 2.67	-.05089 -1.71	.028144 5.37	.001776 0.08	.005321 0.02	-.4558 -0.39	2.08404 1.88	6.38355 0.65	.140276 0.42	.52672 0.30
Tenuresq·treatment		-.00199 -1.06		-.00707 -3.14		-.00342 -2.23		-.36054 -0.35		4.09322 0.47		-.00428 -0.00
Obs.	2262	2262	2021	2021	2786	2786	155	155	128	128	170	170
R-sq	0.5954	0.5965	0.6261	0.6285	0.5866	0.5875	0.8135	0.8140	0.8341	0.8349	0.7736	0.7738

Note *Treatment equals one if women have 1, 2 or 1-2 children. Treatment equals zero, if women have no children. Tenure·treatment is the interaction term of tenure and treatment, and Tenuresq·treatment is the interaction term of tenure squared and treatment. t-statistics. These are presented in the 2nd. line of cells in the tables.*

Closing Remarks

Why does Korea invest so much in school/university education while neglecting job related education and training? As a possible answer, I would suggest a hypothesis in the first essay as follows: Korea gives a great deal of weight to school and university education, because the effect of graduating university on wages is much greater than the wage effects of other lower educational levels. Job training has only a marginal influence on wages in Korea. The wage difference due to educational level can thus not be compensated for with any other human capital investment in Korea. Koreans do not recognize job training as an alternative factor in human capital investment.

In the estimation of the effect of school education and job training on wages in Korea, this study found the following results: Compared to employees who only completed senior high school, employees who completed an elementary school education earn 15% less compared to employees who graduated senior high school, employees who completed junior high school earn 13% less. However, junior college graduates earn 9.4% more, and employees with a bachelor's degree 27.5%, people with a master's degree 54% and those with a PhD 80% more by means of the random effects estimation. Estimation by the Heckman 2 step method demonstrates that the wage level of those who completed elementary school is 36% lower than the wage level associated with senior high school completion. Junior college graduation results in a 17.4% higher wage level, bachelor's degree a 30%, master's degree a 71% and doctoral degree an 89% higher wage level compared to the wage level of senior high school graduation alone. Compared to these effects of education, job training leads to a 3% wage increase compared to no job training using the random effects method.

Concerning the hypothesis: This study found a 27% to 30% wage difference between individuals who obtained a bachelor's degree at university and those who only completed senior high school. Estimation results demonstrate a clear difference between the wages of those with a university degree and those with a lower educational degree, and further demonstrate the meager influence of job training experience on wages. The training effect on wages is only about 3%. According to these estimation results, therefore, the wage difference between university graduates and employees with lower education levels could not be compensated for by accumulating job training experience in one's working life. So, job training could not be recognized as an alternative to university education, as the (central) hypothesis of the study predicted. The dependency on university education in Korea leads to an excessive competition for university entrance, when considering human capital investment.

The next question dealt with by this thesis was why a university education is considered to be so much more important in Korea than is the case in Germany. Concerning this question, Korean researchers often advance the hypothesis that the earnings difference between university graduates and workers with a lower level of education is larger in Korea than in Germany, where vocational training at school and on-the-job training for adults are well developed. German job training reduces earnings differences between educational levels and relaxes the competitiveness of university entrance. This study estimated the effects of school education and job training on earnings and compared the estimation results for Korea with Germany using the quantile regression method.

According to these estimation results, we were able to confirm that the hypothesis of this study failed to offer a valid prediction concerning the earnings effects of education and job training in Korea and Germany. Compared to senior high school graduation (ISCED-97 level 3), an educational level lower than junior high school (ISCED-97 level 1) leads to a -23% to -33% lower earnings at different quantile points in Korea. Junior high school (ISCED-97 level 2) leads to -12% to -16% lower effects, junior college (ISCED-97 level 5B) to 8%-20% higher earnings effects and university graduation (ISCED-97 level 5A and 6) to 30%-40% positive earnings effects in Korea. Job training results in 3.7%-10% earnings effects at various quantile points compared to no job training. In Germany, an educational level lower than the *Haupt-/Realschulabschluss* (ISCED-97 level 1) results in -14% to -71% lower earnings effects at different quantile points, *Haupt-/Realschulabschluss* (ISCED-97 level 2) -13% to -19% lower effects, advanced vocational qualification (ISCED-97 level 5B) such as *Meister* 12%-16% higher effects and university graduation (ISCED-97 level 5A and 6) results in 48%-52% higher earnings compared to *Berufsabschluss* and/or (*Fach-*)*Hochschulreife* (ISCED-97 level 3 and 4, comparable to the education level of senior high school graduation in Korea). The effects of job training vary between 9% and 17% at various quantile points compared to no job training.

These results demonstrate that the earnings difference between university graduation and *Berufsabschluss* and/or (*Fach-*)*Hochschulreife* in Germany is higher than the comparable difference in Korea – i.e. the earnings difference between university graduation and senior high school graduation, although the effect of job training on earnings in Germany is higher than the comparable effect in Korea. The differences are 51% at median using quantile regression and 52% by the OLS estimation in Germany and 30% (median) und 34% (OLS) in Korea. This repudiates the hypothesis of the second essay. The earnings difference between university graduates and workers with a lower level of education was larger in Germany than in Korea. Well-developed job training in Germany could not compensate for the earnings difference between educational levels to the degree that the hypothesis of this study predicted, although the effect of job training on earnings is higher in Germany than in Korea.

The third essay estimates the impact of childbirth and childcare on the earnings of women in Korea and Germany. According to the study by Mincer und Polachek (1974, 1978), this study advances the hypothesis of the earnings discontinuity of working mother due to having children. This study expected an earnings decrease for working women due to childbirth and childcare in this context. As the first comparative study concerning human capital and women's earnings in Korea and Germany, this study analyzed the effects of education, experience and family characteristics on earnings in both countries while testing the hypothesis of this essay.

According to estimation results obtained using the quantile regression technique, an educational level lower than junior high school graduation results in a -8.4% to -11% lower earnings at the different quantile points compared to senior high school graduation in Korea. Junior college results in 9.5% to 11% higher earnings and university graduation+ 35% to 47% positive earnings effects. Junior high school graduation, however, does not result in a significant effect in Korea. Compared to *Berufsabschluss* and/or *(Fach-)Hochschulreife*, an education level lower than *Haupt-/Realschulabschluss* leads to -84% to -41% lower earnings effects for women at different quantile points. *Haupt-/Realschulabschluss* results in -13% to -4% lower earnings, a higher vocational qualification such as *Meister* 4.6% to 17% higher earnings and a university degree results in 22% to 28% higher earnings in Germany. In Korea, when a comparison is made with 12 years of education, 14 years of education results in 10%-13% higher earnings, 16 years of education 35%-42% and 18 years+ of education 52%-67% positive earnings effects. These effects are 13%-20% for 16 years of education, 27%-49% for 18 years+ of education in Germany. 14 years of education does not lead to a significant earnings effect for women in Germany.

Concerning labor market experience, a women's age leads to significant earnings effects in Germany, while age is not associated with any significant result in Korea. The effects of tenure on earnings seem stronger in Korea compared to Germany. However, women's tenure in Korea averages 4 years and its median is only 2.3 years. This may limit the tenure effects on earnings for Korean women. Family responsibility, which has an influence on women's attachment to their jobs, is measured by indicators concerning female household head and single motherhood. Female household heads earn 7% to 9% more at various quantile points than other women in Korea and 3% to 4% more in Germany. Single mothers earn 9% to 22% more than other women in Korea and 7% to 24% more in Germany. The number of children, which may disturb a woman's career progression, results in significant negative effects in Germany. However, this has an insignificant or even positive effect on women's earnings in Korea.

Regarding the primary hypothesis of this essay, estimation results obtained by means of regression discontinuity design indicate the strong negative earnings effect (-65%) of having one or two children for women aged between 30 and 40 in Germany compared to the effect of having no children. A discontinuity in the earnings of women who have one or two children is clearly visible in the tenure-earnings curve for Germany, which verifies the hypothesis. This discontinuity could be also found for the earnings of Korean women aged between 40 and 55, who have one or two children. However, the earnings effect of having one or two children in Korea is unexpectedly positive (37%) unlike the theory of Mincer und Polachek (1974, 1978). This unconventional result could be caused by the high reservation wages of Korean mothers due to the need to care for their children.

While university education in Korea results in 27% to 30% higher wages compared to senior high school graduation, job training experience, which one has accumulated during one's whole working life, only leads to a 3% wage increase compared to no job training experience. With the marginal wage effect of training, compensation or a noticeable reduction of earnings differences between holders of a university degree and those with lower educational levels was unexpected. Thus, job training in Korea is not considered to be a worthy investment opportunity for human capital. University education is thus regarded as a unique human capital, which certainly enhances the earnings of working people in the framework of human capital investment. This inevitably leads to intensifying competition for a university place. In order to reduce this problematically high competitiveness for university entrance in Korea, researchers and politicians have begun to study the German vocational training system. It is often assumed that the well-developed German dual system leads to a reduction of earnings differences between holders of university degrees and senior high school degrees or lower, and thereby results in a balanced distribution of pupils between vocational oriented and academic oriented schooling routes. As a consequence of this, competition for university entrance is not so problematic in Germany. According to the estimation of this study, however, earnings differences between university graduates and those who left education at the upper secondary level is higher in Germany than in Korea, although job training in Germany impacts more strongly on earnings than in Korea. So, this result rejects the assumption made by the Koreans. What is behind the balanced distribution of pupils between vocational oriented and academic oriented schooling routes and the reduction in competition for university entrance in Germany? This question still remains unsolved in Korea.

Not only Korea but also Germany seriously suffers from a low fertility rate. This study demonstrates how significant childbirth and childcare is for the careers/earnings of women. According to the estimate of this study, a working mother's career was significantly interrupted due to having children in Korea and Germany. Young working mothers aged between 30 and 40 and with one or two children earn about 65% less compared to women without children in Germany. In this situation, a decision in favor of children would not be easy for an ambitious woman. In Korea, the difficulties of childcare for women seem not to be reflected in earnings levels, but rather in their working experience and tenure. The estimation result shows us that the labor market experience of Korean women does not have any significant effect on earnings. Their average tenure is only about 4 years and the median tenure is 2.3 years. However, the earnings effects associated with having one or two children were unexpectedly positive (37%) for Korean working mothers aged between 40 and 55 compared to working women without children. At a guess: The lack of family friendly policies offered by firms and the government, the high responsibility of women for childcare, and the difficulties of school education for Korean children all serve to decrease women's chances to work and increase their reservation wages.

The biased behavior of Koreans concerning human capital investment – too high a desire to go on to university – leads to an oversupply of university graduates in the labor market and subsequently leads to a distortion between labor demand and -supply. Women might also suffer from this biased behavior in Korea, where mothers neglect their educational and vocational qualifications in favor of childcare. The high desire of Koreans to go on to university could be caused by the lack of a profitable alternative to a university education in terms of human capital theory. In order to balance the distribution of pupils between a vocational and academic educational route, the exclusive position accorded to a university education and the meager profitability of job training should be renegotiated. It would be desirable to consider how to enhance the productivity of vocational education and training in Korea. Although an improvement in vocational education and training and the introduction of recognized qualifications would not automatically lead to a reduction in earnings differences between a higher and a lower educational level, as the German case demonstrated, such an enhancement of the job training system and related programs could increase the economic effectiveness of job training in general, along with any earnings effects, which would result in more interest in vocational education and training by pupils and their parents. From this point of view, taking as an example of the German dual system and further related training, which politicians and researchers have undertaken in Korea, seems a reasonable first step in balancing the biased behavior of Koreans concerning human capital investment and dealing with the disparity in the labor supply and further demands concerning education. This would also reduce the burdens on women and families concerning the education of their children, which would ease the decision to have more children and help mothers to become more active in the labor market.

References

Ashenfelter, O. and Zimmerman, D.J. (1993), "Estimates of the Returns to Schooling from Sibling Data: Fathers, Sons, And Brothers," NBER Working Papers, 4491, National Bureau of Economic Research.

Ashenfelter, O. and Zimmerman, D.J. (1997), "Estimates of the Returns to Schooling from Sibling Data: Fathers, Sons, And Brothers," *Review of Economics and Statistics*, 79(1), pp. 1-9.

Becker, G.S. (1962) "Investment in Human Capital: A Theoretical Analysis," *Journal of Political Economy*, LXX, pp. 9-49.

Becker, G.S. (1964) *Human Capital*, New York: Columbia University Press.

Heckman, J.J. (1979), "Sample Selection Bias as a Specification Error," *Econometrica*, 47(1), pp. 153-161.

Hippach-Schneider, U., Krause, M. and Woll, C. (2007), *Berufsbildung in Deutschland – Kurzbeschreibung*, CEDEFOP(European Centre for the Development of Vocational Training) Panorama Series, 136, Luxemburg: Amt für amtliche Veröffentlichungen der Europäischen Gemeinschaften.

Kim, K.H. (2006), "The Vocational Education System and the Priority given to Talent and Attitude in Education," *Eduforum*, 124, pp. 17-19 (in Korean).

Koenker, R. (2005), *Quantile Regression*, Cambridge: Cambridge University Press.

Koenker, R. and Bassett, G. (1978), "Regression Quantiles," *Econometrica*, 46 (1), pp. 33-50.

Korean Educational Development Institute (2005), *Educational Indicators for Primary and Secondary Levels of Schooling*, RM2005-56, Seoul (in Korean).

Korean Educational Development Institute (2009), *2008 Employment Statistics*, http://cesi.kedi.re.kr/index.jsp, (2009.10.15), Seoul (in Korean)

Korean Statistical Information Service (2010), *The Number of Unemployed People according to Age and Education Level*, http://kosis.kr/nsportal/themes/themes _04List.jsp, (2010.06.12), Seoul (in Korean).

Mincer, J. (1974), *Schooling, Experience and Earnings*, New York: Columbia University Press.

Mincer, J. and Polachek, S. (1974), "Family Investments in Human Capital: Earnings of Women," *Journal of Political Economy*, 82(2), pp. 76-108.

Mincer, J. and Polachek, S. (1978), "Women's Earnings Reexamined," *Journal of Human Resources*, 13(1), pp. 118-134.

OECD (1999), *Classifying Educational Programmes – Manual for ISCED-97 Implementation in OECD Countries*, Paris.

OECD (2007), *Babies and Bosses – Reconciling Work and Family Life (Vol. 5): A Synthesis of Findings for OECD Countries*, Paris.

OECD (2009a), *Education at a Glance 2009: OECD Indicators*, Paris.

OECD (2009b), *Society at a Glance 2009: OECD Indicators*, Paris.

Thistlewaithe, D.L. and Campbell, D.T. (1960), "Regression-Discontinuity Analysis: An Alternative to the Ex Post Facto Experiment," *Journal of Educational Psychology*, LI, pp. 309-317.